ERRATUM

The following information was omitted from the copyright notice of this volume when published:

Fundamentals of Laser Optics
Kenichi Iga

0-306-44604-9
Plenum Press, New York, 1994

FUNDAMENTALS OF LASER OPTICS

LASERS, PHOTONICS, AND ELECTRO-OPTICS

Series Editor: H. Kogelnik

FUNDAMENTALS OF LASER OPTICS
Kenichi Iga

FUNDAMENTALS OF LASER OPTICS

KENICHI IGA
Tokyo Institute of Technology
Yokohama, Japan

TECHNICAL EDITOR
RICHARD B. MILES
Princeton University
Princeton, New Jersey

SPRINGER SCIENCE+BUSINESS MEDIA, LLC

Library of Congress Cataloging-in-Publication Data

Iga, Ken'ichi, 1940-
Fundamentals of laser optics / Kenichi Iga ; technical editor, Richard B. Miles.
p. cm. -- (Lasers, photonics, and electro-optics)
Includes bibliographical references and index.

1. Lasers. 2. Electrooptics. I. Miles, Richard B. (Richard Bryant), 1948- . II. Title. III. Series.
TA1675.I33 1994
621.36'6--dc20 93-50100
CIP

ISBN 978-1-4613-6057-5 ISBN 978-1-4615-2482-3 (eBook)
DOI 10.1007/978-1-4615-2482-3

Originally published by Plenum Press, New York in 1994
Softcover reprint of the hardcover 1st edition 1994

PREFACE

The laser, initially called the "optical maser," was proposed in 1958 by Charles Townes and Arthur Schawlow; in 1960, Theodore Maiman was the first among several researchers to achieve laser oscillation by using a ruby crystal. In the following quarter of a century, a considerable amount of research and development has taken place, and the laser is now utilized for many diverse applications, ranging from the commonplace compact disk to intricate surgical applications in medicine. Since I first entered the laboratory of Professor Yasuharu Suematsu in 1962 to complete my thesis, I have been studying the new field of laser optics. In spite of many expectations and a vast investment in research, the first practical use of lasers was difficult to achieve. The late Professor K. H. Zchauer of Univ. Erlangen once jokingly remarked that *laser* was defined by an English physicist as "Less Application of Stimulated Expensive Research." In a similiar vein, Dr. Herwig Kogelnik reminded me that in the early 1960s, *maser* was often called "Money Acquisition Scheme for Expensive Research."

Initially I worked with a ruby laser, then with a helium–neon-gas laser, and am presently engaged in semiconductor laser research. There are probably not a large number of researchers who have had the opportunity to build these three representative types of lasers. My primary objective of study lies in optical communications however, and therefore, I have been approaching the laser mainly as a lightwave propagator. In 1973, a graduate school was established at the Nagatsuta Campus of the Tokyo Institute of Technology, where I had been working. It was at that time that I started to give a new lecture series on laser optics, which covered the basic principles of the applications of lasers. This lecture series was presented to graduate students in the Information Processing and Applied Electronics Departments (not all of whom were necessarily doing laser research). I have revised my lecture materials each year since; reviewing and compiling them into textbook form, I decided to publish it under the title of "Fundamentals of Laser Optics."

The title of this book implies that priority has been given to the fundamentals of the laser. The purpose of this book is to describe the principles of lasers and the behavior of laser beams optically. It is these fundamental concepts that are required in applications for the utilization of lasers. In other words, the goal is to introduce the basic concepts of lasers that need to be understood for industrial laser use.

In Chapter 1, the basic concept of lasers is surveyed. In Chapter 2, the extensive field of laser applications is considered. In Chapters 3 and 4, gas, liquid, and solid-state lasers are briefly introduced. Chapter 5 describes semiconductor lasers, including their fabrication and their operational characteristics. In Chapters 6 and 7, fundamental principles of light beams and optical waveguides are discussed. In Chapter 8, laser resonators and resonator modes are considered. In Chapter 9, equations describing the basic behavior of lasers are developed using a density matrix formalism. In Chapter 10, these equations are then used to derive rate equations that describe the basic characteristics of lasers. In Chapter 11, fundamentals of nonlinearity and saturation in lasers are described. In Chapter 12, some aspects of modulation and the generation of short optical pulses are studied. In Chapter 13, various concepts of noise in a laser oscillator, including intensity noise, phase noise, etc., are introduced. In Chapter 14, advanced technology for semiconductor lasers and the integration of semiconductor lasers into large-scale integrated optical circuits are discussed. Lastly, in Chapter 15, we introduce a new type of semiconductor laser, i.e., surface-emitting laser.

This book is written for students at the junior and senior level of science and engineering schools, for graduate students, and for researchers who plan to be involved with the optoelectronics field.

ACKNOWLEDGMENTS

I would like to thank Dr. Herwig Kogelnik of AT&T Bell Laboratories (the editor of this series), for suggesting that I write the book, and, moreover, I would like to thank him for reading the initial manuscript and for giving much advice that greatly improved the contents. Without his sizable effort and dedicated encouragement, I might have given up on the completion of the book. I would also like to thank Professor Richard B. Miles for his helpful comments as well as for editing the language and contents of this book.

Also, I gratefully acknowledge the continuing encouragement of Emeritus Professor Yasuharu Suematsu, the former president of the Tokyo Institute of Technology. I would also like to thank Emeritus Professor Koichi Shimoda of the University of Tokyo and Emeritus Professor Toshiharu Tako of the Tokyo Institute of Technology for their helpful advice. I also thank the following colleagues who provided valuable information related to laser optics: Professor Motoichi Ohtsu, who has also given this course, Professor Kazuhito Furuya of the Electrical and Electronics Department, Professor Yasuo Kokubun of the Yokohama National University, Professor Masahiro Asada, and Professor Fumio Koyama. I would also like to thank the staff of Ohmsha, Ltd., who gave additional editorial help.

CONTENTS

Chapter 13. Laser Noise

Chapter 14. Advanced Technology for Semiconductor Laser Fabrication and Integration

Chapter 15 Surface-Emitting Lasers

ONE

THE BASIC CONCEPT OF LASERS

This chapter serves as an outline of various important aspects of lasers such as their history, the coherence of laser beams, and the characteristics of lasers. Industrial fields where lasers and light are important components will be introduced in the following chapter.

1.1. WHAT IS A LASER?

"Laser," the subject of this book, is an acronym of

*L*ight
*A*mplification by
*S*timulated
*E*mission of
*R*adiation

It was originally used to describe light amplification, but has more recently come to mean an optical oscillator which implies two functions, namely, light amplification by stimulated emission and feedback of light by a reflecting mirror. Therefore, the word *laser* now denotes an oscillator unless otherwise specified.*

* The verb *lase* is derived from the word *laser* and may be used, for example, as in a "lasing" diode.

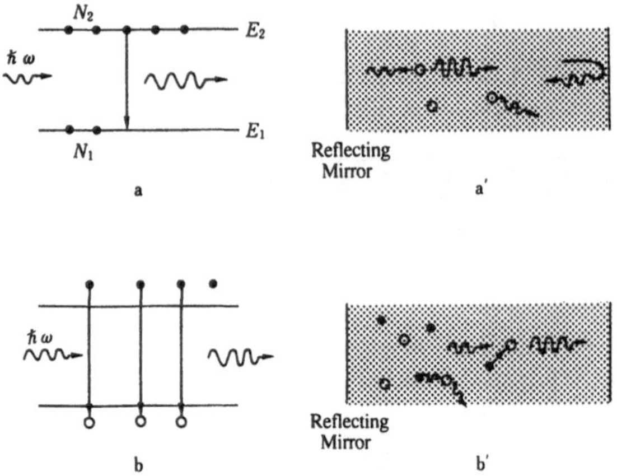

Figure 1.1. (a) Two-level laser; (b) semiconductor laser.

As shown in Fig. 1.1, either a molecule or a semiconductor may emit light with a wavelength which is inversely proportional to the energy difference between the energy states E_1 and E_2. When the energy difference $\Delta E = E_2 - E_1$ is expressed in units of electron volts (eV), the wavelength, λ_0, can be written as:

$$\lambda_0 = 1.2398/\Delta E \qquad (\mu\text{m}) \tag{1.1}$$

Since the wavelength depends on the energy level difference, it is characteristic of the type of atom, molecule, or semiconducting material used as a laser medium. In practice, amplification by stimulated emission occurs with some finite spectral width centered on λ_0. The width is on the order of 0.05 Å in a gas laser, and can be as wide as 10 to 100 Å in a solid-state laser. In a semiconductor, the electron energy has a broad band structure, as illustrated in Fig. 1.1(b), and the gain width can be as large as 400 Å. In a compound semiconductor, the wavelength of the emitted light can be changed over a wide range by changing the material composition.

A laser oscillator, just like an oscillator in electronics, requires a resonator in which the feedback of light occurs. For this purpose, a Fabry–Perot resonator, consisting of two reflecting mirrors placed face to face, is most often employed. Figure 1.2 shows the consequences of using such a resonator configuration. The light is reflected and confined between the two reflecting mirrors so that it produces a standing wave. Stimulated emission in the resonator cavity occurs with the same phase as that of the standing wave. The resonant spectrum will be mentioned later and given by Eq. (1.3).

Assume that the stimulated emission gain obtained when light propa-

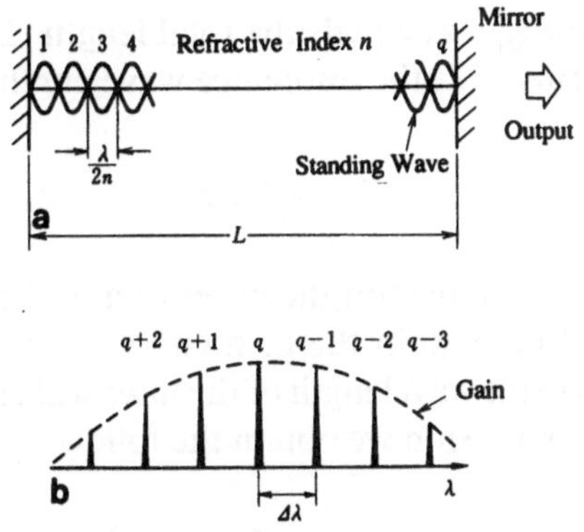

Figure 1.2. Standing wave and spectrum of a laser resonator. (a) Standing wave of a longitudinal mode; (b) resonant spectrum of a longitudinal mode.

gates through 1 cm of the active medium is g (cm^{-1}), and that the loss over the same distance is α (cm^{-1}). Further assume that the reflectivity of the mirrors at either end of the cavity is R_1 and R_2, and the cavity length is L. Then the condition that the intensity of light be constant after a single round trip yields a threshold gain, g, given by:

$$g = \alpha + (1/2L)\ln(1/R_1R_2) \tag{1.2}$$

For a gas laser, the absorption loss, α, is small enough to be ignored when compared with mirror losses. Typically, we have $L = 10$ cm, $R_1 = R_2 = 0.98$, and $g = 2 \times 10^{-3}\ cm^{-1}$. In the case of a semiconductor, light loss caused by free electron absorption occurs and α is nearly 10 cm^{-1}. The length of a semiconductor resonator, L, is on the order of 0.03 cm, and the reflectivities are $R_1 = R_2 = 0.3$, due to Fresnel reflection between the semiconductor (refractive index = 3.5) and air. These numbers give a value of g of approximately 50 cm^{-1} required for threshold.

If the amplification gain of the light is greater than the total loss, i.e., the sum of absorption loss and reflection loss, and, furthermore, if a resonance wavelength exists within the gain bandwidth, then oscillation occurs. A laser beam is generated and the spectral linewidth of the beam becomes significantly narrower than that of luminescence or spontaneous emission associated with the same transition.

The Fabry–Perot resonator will oscillate with any optical field whose wavelength has an integral number of half wavelengths between the two mirrors. Let us assume that n is the refractive index of the medium in the resonator as shown in Fig. 1.2. The wavelength of light in the medium is then λ/n, where λ is the light wavelength in free space. If this wavelength is resonant, then $\lambda/2n$ times an integer, q, equals the total length, L, of the Fabry–Perot resonator, and, in terms of L, the resonance wavelength is given as follows:

$$\lambda = 2nL/q \tag{1.3}$$

Usually, the value of q is large (on the order of a million), so the next resonator mode may still fall within the gain curve of the lasing medium. If q changes by 1, the resonant wavelength of the laser will change by the amount $\Delta\lambda$. From the above expression we obtain the following equation:

$$|\Delta\lambda| \cong \lambda^2/2nL \tag{1.4}$$

When the gain, g, at the wavelength $\lambda + \Delta\lambda$ reaches threshold, oscillation is possible at this second wavelength. In all but the smallest lasers, the gain bandwidth of the laser includes many modes and these modes with their various wavelengths are called longitudinal modes.

In the direction perpendicular to light propagation between the two mirrors of the Fabry–Perot resonator, the light may have a beam shape with various forms which are called transverse modes. For a wide range of applications, it is desirable that the transverse modes be simple. The approximate wavelength of the laser output is determined by the resonant transition of the atoms or molecules which are used, but the specific details of the output depend on the laser resonator. In lasers with a wide gain spectrum, many longitudinal modes are possible and wavelength tuning of those modes can be achieved by varying the resonance wavelength of the resonator.

The laser output power is determined by the excitation mechanism and the gain reduction at high intensities due to the circulating laser light. Figure 1.3 shows an outline of the wavelengths and output powers of various lasers. The gain bandwidths of gas lasers and solid-state lasers are typically very narrow, whereas those of semiconductor lasers and dye lasers are broad.

1.2. HISTORY OF LASERS

1.2.1. Development of Masers

Table 1.1 shows the development history of the maser (*m*icrowave *a*mplification by *s*timulated *e*mission of *r*adiation). Research on masers was motivated by the idea that a stable frequency source is produced by utilizing a transition between the energy levels of an atom or a molecule. In about 1951,

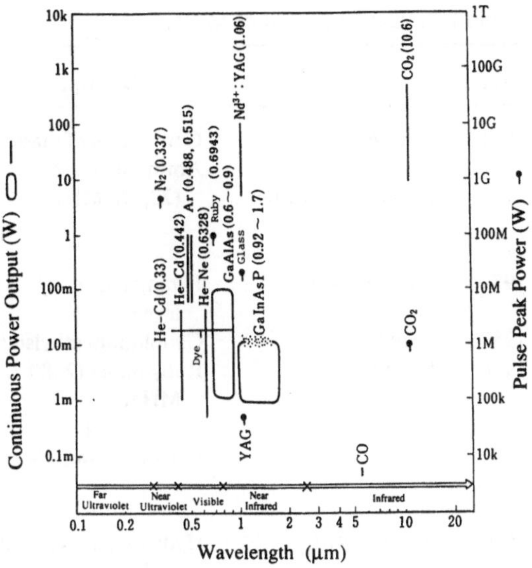

Figure 1.3. Wavelength and output of various lasers. (——) and □ denote a continuous laser wave output shown on the axis of ordinates on the left and —• denotes a pulse front value on the axis of ordinates on the right.)

Townes and Schawlow, and Basov and Prokhorov, independently conceived of masers on the basis of such a principle.[1] A 23,870-MHz ammonia maser was utilized for some time as a frequency standard. After that, a ruby maser was put to practical use as a low-noise amplifier for a high-sensitivity radar or radio telescope. A hydrogen maser has achieved the highest degree of frequency stability using a resonant level of hydrogen at 1420 MHz. On the other hand, rubidium (Rb) masers using rubidium metal vapor were studied as a compact frequency standard, and they are once again drawing attention because new semiconductor lasers are now able to be utilized for efficient optical pumping. In any case, masers were soon followed by lasers, which now rank as the highest-performance devices for frequency standards.

1.2.2. Development of Lasers

Several lasers were proposed from 1958 through 1960. However, the first laser to emit coherent light was the ruby laser invented in 1960 by Maiman.[2] Table 1.2 shows important milestones in the development of lasers. The helium–neon-gas laser was developed by Javan *et al.*[3] in 1960, and semiconductor lasers were realized in 1962 by four groups in the United States.[4–7] Since the laser was first introduced in the early 1960s, a wide variety of lasers have been developed, and it can now be said that the era of the laser has come at last. Various devices, equipment, and systems utilizing lasers have been introduced into practical use, and reports on continuing development arouse

Table 1.1. Development of Masers

Year	Name	Description	Applications
1951	Townes	Conception of maser	
1954	Gordon Zeiger, and Townes	Ammonia maser (23,870 MHz)	Microwave oscillation ↓ Atomic clock Accuracy of frequency 10^{-10}
1956	Bloembergen	Paramagnetic crystal (3 levels)	
1957	Scovil	Gadolinium ethylsulfate	
1958	Makhov	Ruby maser (2380 MHz) ↓ Low-noise amplifier	High-sensitivity radar Radio telescope Space communication
1960	Ramsey *et al.*	Hydrogen maser (1420 MHz)	Frequency standard Accuracy of frequency 10^{-13} to 10^{-14}
1964	Davidovitz	Rubudium maser (6835 MHz)	Accuracy of frequency 10^{-12}

much interest at the Conference on Lasers and Electro-Optics (CLEO), the largest international conference in this field.

1.3. CHARACTERISTICS OF LASER BEAMS

Coherence represents a remarkable characteristic of the laser beam. There are two aspects of coherence: temporal coherence and spatial coherence. The former represents monochromaticity (or single frequency), and the latter, monodirectionality. Though the laser beam is said to be a coherent light source, it is not completely coherent. Because the emitted light or wave packet is finite in length and has fluctuations in amplitude and phase, it has a spectral width greater than zero. This width is not the width of the gain curve, due to gain narrowing, as described later, but it is a narrow spectral width around one or more cavity resonances, the magnitude of which relates to noise in the laser medium and oscillator cavity.

When the spectral width is $\Delta\lambda_c$, the maximum path-length difference over which the beam is still capable of generating an interference pattern, as shown in Fig. 1.4, is:

Table 1.2. History of Lasers

1958	Schawlow, Townes	Proposal of optical maser K light pump, unsuccessful
	Basov	Theory of optical maser
1959	Schawlow	Proposal of ruby laser
1960	Maiman	Ruby laser oscillation successful; $\lambda = 0.6943\ \mu m$ (3 levels)
	Sorokin	Uranium laser, $\lambda = 2.5\ \mu m$ (4 levels)
1958–1960	Aigrain (France) Nishizawa (Japan) Basov (USSR)	Proposal of semiconductor laser
1960	Javan *et al.*	He–Ne laser, $\lambda = 1.15\ \mu m$ (internal reflecting mirror)
1961	Rigrod	He–Ne laser, $\lambda = 1.15\ \mu m$ (external reflecting mirror)
1962	White	He–Ne laser, $\lambda = 0.6328\ \mu m$
1962	Nathan *et al.* (IBM) Hall *et al.* (GE) Quist *et al.* (MIT) Holonyak *et al.* (Ill. Univ.)	Semiconductor laser
1963	Mathias	N_2 laser
1964	Geusic	YAG laser
	Bridges	Ar-ion laser
	Patel	CO_2
1965	Khokhlov	KDP optical parametric oscillator
	Wang	ADP optical parametric oscillator
	Giordmaine	$LiNbO_3$ optical parametric oscillator
1966	Sorokin	Dye laser
1969	Hayashi, Panish	GaAs/GaAlAs, double heterostructure semiconductor laser
1970		Double heterostructure room temperature CW
1973		Coherent ultraviolet ray
		Generation of x ray
1975		Elongation of life of semiconductor laser (10,000 hr)
1976	Hsieh	GaInAsP semiconductor laser ($\lambda \cong 1.1\ \mu m$)
1978	Several groups	GaInAsP semiconductor laser ($\lambda \cong 1.3\ \mu m$)

$$l_c = \lambda^2/\pi\Delta\lambda_c \tag{1.5}$$

l_c is called the coherence length and ranges from 10 to 100 cm in semiconductor lasers, and sometimes reaches several kilometers in stabilized gas lasers. l_c is a measure of the temporal coherence or, equivalently, the monochromaticity of the laser beam. Long-distance interference and coherent optical communications utilizing such phase information can be regarded as typical utilization of the temporal coherence of laser beams.[8]

The power density per unit wavelength of the laser depends on the spectral width. For example, the power density of a 1-kW white light source with

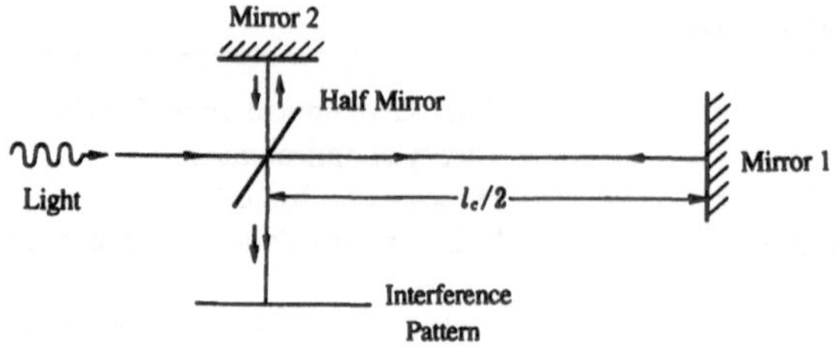

Figure 1.4. Temporal coherence measurement of light by using an interferometer.

a spectral width of approximately 0.4 μm is 25 mW/Å, while that of a 1-kW laser with a spectral width of 0.1 Å is 10 kW/Å. High spectral power density per unit wavelength means that the electric field component is large over a narrow wavelength range. This characteristic of lasers can be utilized for the study of high-electric-field phenomena.

Spatial coherence is a measure of the transverse distance across the beam over which a constant phase relationship exists. Spatial coherence can be perceived intuitively by observing various interference patterns. The length in the transverse direction (perpendicular to the direction of propagation) over which the phase front is constant is called the coherence length, x_c. A typical laser beam has a coherence length of approximately 1 mm as it exits the laser, and, if broadened by a high-quality lens, the coherence length can be made much larger. However, it cannot be made extremely large because of aberrations or inhomogeneities in the lens. If a laser beam with a coherence length x_c is focused by a lens, as shown in Fig. 1.5, the diameter D of the beam spot is given as:

$$\Delta D \cong 2.44\lambda f/x_c \tag{1.6}$$

where x_c must be smaller than the diameter of the lens. If $f/D = 1$, $\lambda = 0.78$

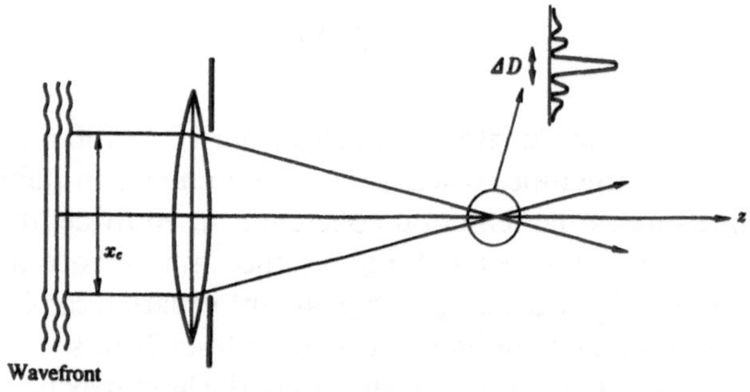

Figure 1.5. Spatial coherence and focusing spot.

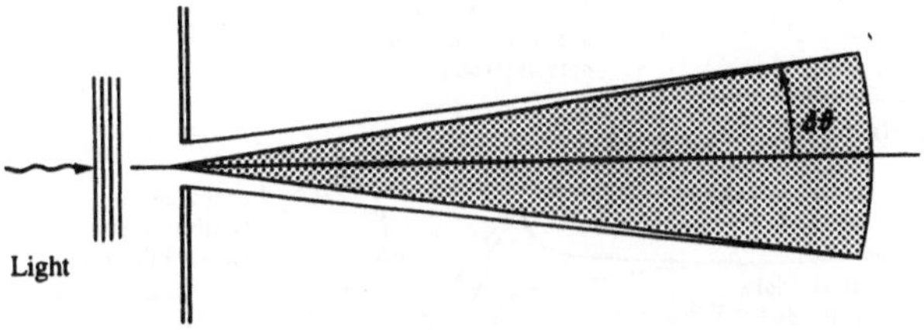

Figure 1.6. Light propagation and broadening angle.

μm, and $x_c = D$, the diameter of the beam spot $\Delta D = 0.7\ \mu$m. This spot size is called the diffraction limit and the ability to focus a laser beam to that limit is one of the features of spatial or transverse coherence.

In a similar manner, the divergence angle, $2\Delta\theta$, of an emitted beam into space is related to the coherence length, x_c, by the expression (see Fig. 1.6):

$$2\Delta\theta \cong 1.22\lambda/x_c \tag{1.7}$$

Laser beams can be made highly directional because a large value of x_c may be obtained. This is usually considered a characteristic feature of a laser.

1.4. FIELD OF OPTOELECTRONICS

The application of lasers will be detailed in Chapter 2. In this section, we illustrate the fields where lasers are utilized with a laser tree, shown in Fig. 1.7.

The application fields of lasers in optoelectronics may be classified into lightweight optics and heavyweight optics for convenience. Lightweight optics includes optical communications, laser disks, optoelectronics equipment, etc., which utilize the information of light effectively, and lasers play the role of a device which corresponds to a transistor in electronics. On the other hand, heavyweight optics utilizes solid-state or gas lasers with high energy or high power output as large-sized equipment for energy development, medical uses, and industrial processing. The applications of heavyweight optics is surely expected to grow with large-scale projects. However, lightweight optics is growing into a far-larger-scale industry by utilizing semiconductor lasers. The author has primarily been studying lightweight optics and will describe in this book the fundamentals of solid-state and gas lasers, as well as Fabry–Perot resonators with plane mirrors. The reason for emphasizing these devices is as follows: A basic understanding of the concepts of mode control in a solid-state laser resonator is important for the surface-emitting

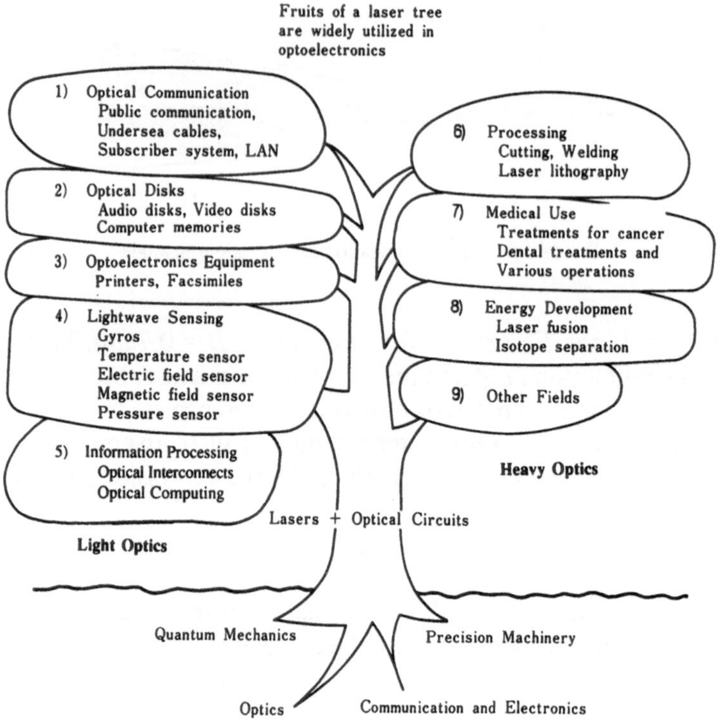

Figure 1.7. Laser tree.

semiconductor laser that we will be studying in Chapters 4 and 5. In addition, the concept of light beams is common to all kinds of lasers. Therefore, to prepare for writing this book, the author investigated solid-state and gas lasers not only because he wanted to include all kinds of lasers, but also because he formed a clear view that the principal thrust of the entire volume is the study of lightweight optics.

PROBLEMS

1.1. Show that the photon wavelength λ_0 corresponding to the energy difference ΔE is expressed by Eq. (1.1).

1.2. Show that the gain necessary for balancing the loss in cavity is expressed by Eq. (1.2). If we assume that the cavity length $L = 100\ \mu$m and $R_1 = R_2 = 0.7$, obtain the threshold gain gth.

1.3. What is Fresnel reflection? Explain how its reflectivity is given in terms of refractive index. Calculate the reflectivity of a semiconductor surface whose refractive index is 0.35.

1.4. Prove Eq. (1.4).
1.5. Prove Eq. (1.5). Refer to some textbook on optics.
1.6. Prove Eq. (1.6).
1.7. Prove Eq. (1.7). Obtain $2\Delta\theta$ for $\lambda = 1.3\ \mu m$ and $x_c = 2\ \mu m$.
1.8. Explain in 50 words the meaning of coherence.

REFERENCES

1. A. L. Schawlow and C. H. Townes, *Phys. Rev.* **112,** 1940 (Dec. 1958).
2. T. H. Maiman, *Nature* **187,** 493 (1960).
3. A. Javan, W. R. Bennett, Jr., and D. R. Herriot, *Phys. Rev. Lett.* **6,** 106 (Feb. 1961).
4. M. I. Nathan, W. P. Dumke, G. Burns, F. H. Dill, Jr., and G. Lasher, *Appl. Phys. Lett.* **1,** 62 (Feb. 1, 1962).
5. R. N. Hall, G. H. Fenner, J. D. Kingsley, T. J. Soltys, and R. D. Carlson, *Phys. Rev. Lett.* **9,** 366 (1962).
6. T. M. Quist, R. H. Rediker, R. J. Keyes, W. E. Krag, B. Lax, A. L. McWhorter, and H. J. Zeiger, *Appl. Phys. Lett.* **1,** 91 (Feb. 15, 1962).
7. N. Holonyak, Jr., and S. F. Bevacqua, *Appl. Phys. Lett.* **1,** 82 (Feb. 1962).
8. Y. Suematsu and K. Iga, *Introduction to Optical Fiber Communication,* rev. 3rd ed., 3rd printing, Ohmsha, Tokyo (1992).

TWO

LASER APPLICATIONS

This chapter will first review the applications of lasers. The optoelectronics field regarding laser technology includes optical fiber communications, optoelectronic equipment, lightwave sensing, medical optics, various laser processes, and so on. These systems are based on new concepts of system design. Here we attempt to classify laser characteristics and the applications of lasers. Next, as some representative applications, we overview optical communications, optoelectronic equipment, and medical systems.

2.1. LASER CHARACTERISTICS AND APPLICATION AREAS

This book will describe the important characteristics of various lasers. Before going into details, areas of technical application of lasers will be discussed. Table 2.1 gives examples of laser applications. Industrial fields such as optical communications and, more generally, optoelectronics, are being aggressively developed. In addition to these areas, some special devices and equipment which provide new capabilities have been established in heavy industries and in measurement and medical fields. Some of the application systems which are already in use, or which will be of use, are highlighted. Associated lasers, wavelengths, and typical output powers are shown in Table 2.1, along with the characteristics of the laser which are particularly relevant for that application.

Table 2.1. Applications of Lasers

Field	Applicable equipment	Characteristics[a]	Laser	Wavelength (μm)	Output
Optical communication (fiber, air, space)	Transmitter, Repeater	S, D	GaAlAs	~0.85	10 mW
			GaInAsP	1–1.6	10 mW
Photoelectric devices	Optical disk	F	He–Ne	0.633	~5 mW
			GaAlAs	0.78	~5 mW
	Printer	F, E	CO_2	10.6, 0.63	
	Facsimile		He–Ne		
	Register	F	He–Ne	0.63	~5 mW
	Holography	C	Ar	0.48	
	Special photo		He–Ne	0.63	
	Display	F	Ar, Kr	Visible	Several W
	Read-in device	C	He–Ne	0.63	~5 mW
Laser industry	Holing	F, E	CO_2	10.6	Several tens of W
	Scriber	F, E	YAG	1.06	~30 W
	Board cutter	E	CO_2	10.6	Several tens of W
	Frequency matching	F, E	Ruby	0.69	~MW (pulse)
	Microelectronics optical source	C	He–Cd	0.33	~10 mW
	Nuclear fusion	E, F	Glass	1.06	~TW (pulse)
			CO_2	10.6	
	Isotope separation	S, E	Chemical	16, etc.	
	Fiber elongation	E	CO_2	10.6	Several tens of W
	Purity evaporation	E	CO_2	10.6	Several tens of W

Measurement	Spectroscopy	S	Dye	Visible	
			Parametric oscillator	Infrared	
	Distance and length	C	He–Ne	0.63	10 mW
			Ruby	0.69	~MW (pulse)
	Radar	D	Ruby	0.69	~MW (pulse)
	Air pollution	S	CO_2	10.6	
			Dye	Visible	
	Laser sight	D	He–Ne	0.63	10 mW
	Physical experiment, physical measurement	C, D	He–Ne	0.63	~5 mW
	Special photo	D, C	Ruby	0.69	~5 mW (pulse)
	Speed measurement	D	He–Ne	0.63	10 mW
	Astronomical observation	D	Ruby	0.69	~MW (pulse)
Medical engineering	Surgery, dermatology	F, E	Ruby	0.69	~MW (pulse)
			YAG	1.06	~MW (pulse)
	Ophthalmology (optical solidification)	E	Ruby	0.6943	0.5–1.5 J
			Ar	0.488, etc.	0.5–2 W
	Dentistry	E	Ruby (QSW)	0.6943	10–30 J

[a] Abbreviations: C, coherency; D, directivity; E, energy density; F, focusing; S, single wavelength.

2.2. OPTICAL COMMUNICATIONS

2.2.1. Optical Communication Systems

The optical communication system is one using a lightwave as a transmission carrier.[1] In most of the current systems, it is common to transmit light with intensity modulation and direct defection (IMDD) by using optical fibers and pulse code modulation (PCM). In such systems, even though the wavelength of the light source, such as a semiconductor laser or an LED, might change because of temperature variation, there is almost no influence on the quality of the transmission. This characteristic of IMDD with PCM is a great advantage. Regenerative repeaters are used at intervals of every 20–50 km in long-haul telephone systems to compensate for the deterioration of signals. By comparison, the repeaters in the conventional coaxial cables must be provided at intervals of 1 to 2 km. This is why optical fibers have a wide range of application for public communication systems all over the world. Optical fiber networks are also utilized for short distances as local area networks. These are particularly desirable in electronically noisy places such as factories or near railways, since optical fibers are immune to electromagnetic interference.

Most of the telephone and telegraph companies in the world are planning to develop the integrated services digital network system using optical fiber transmission lines which will include not only telephone traffic but also video programs. In 1985, 3400-km-long optical cables were installed in Japan between Asahikawa and Kagoshima (see Fig. 2.1). With a single pair of fibers, 5760 voice channels can be operated in two-way service. Twelve pairs of fibers have been laid between Tokyo and Osaka, and at least four pairs in other districts in Japan. In addition, undersea cables have been employed between islands. Optical communication systems are being developed all over the world. The FT3 system of AT&T was the first big system in the United States (900 km between Washington, D.C., and Boston). The telephone repeater system in England, the BIGFON system in Germany, BERLITZ in France, etc., are all operating.

For the undersea optical fiber communication system, optical fibers are laid under the sea between countries or islands. There is little difference between a domestic system and an international system; common features are as follows:

1. Fiber cables are sunk from ships.
2. Cables have high tensile strength, which is required in order to pull them up for repairs.

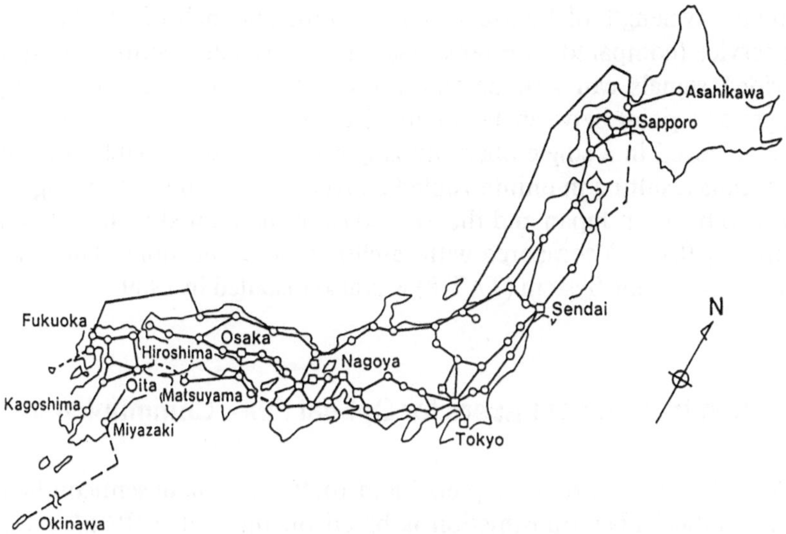

Figure 2.1. Optical transmission system deployed in Japan (as of 1989). Figure courtesy of Dr. Hiroshi Murata.

3. Hydraulic pressure-resisting repeaters are provided at intervals of approximately 50 km.
4. Cable parts must have high reliability, because their maintenance is difficult compared with land-based systems.

The optical fiber undersea cable across the Pacific Ocean was completed in 1989, as shown in Fig. 2.2. Six single-mode fibers with a bit rate of 210 Mbit/

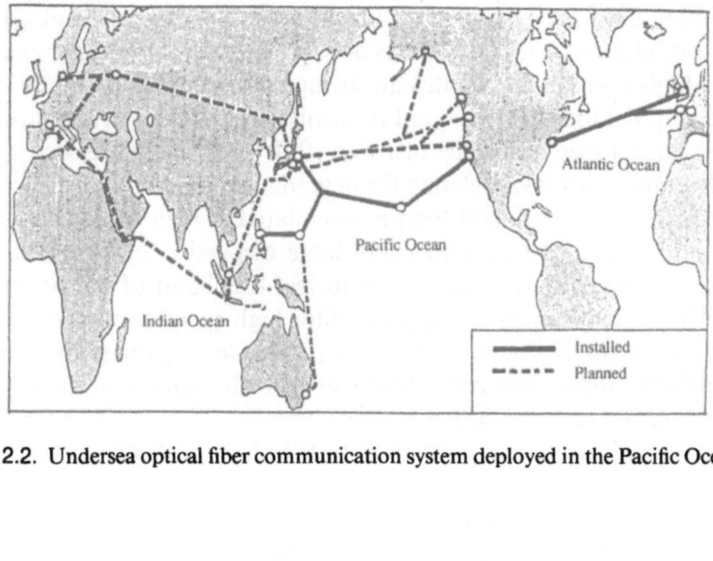

Figure 2.2. Undersea optical fiber communication system deployed in the Pacific Ocean.

sec and a wavelength of 1.3 μm were used; 7560 channels of telephone can be in service (compared with 6000 channels for satellite communication), and video signals can also be transmitted. Note that the Pacific cable branches at a point between Japan and Guam. Also, only one-way service can be achieved in a single fiber; utilizing a pair of fibers enables two-way service, thus resulting in uninterrupted conversations and less time lag (one round-trip between Japan and the west coast of the United States takes approximately 0.1 sec) compared with satellite communications. The Trans-Atlantic Fiber-Optic System (TAT-8) was also installed in 1989.

2.2.2. Semiconductor Lasers for Optical Fiber Communication

As will be discussed in Chapters 5 and 14, the dominant semiconductor laser for optical fiber transmission is based on the GaInAsP/InP system which has a wavelength in the range of 1.3 to 1.6 μm. This quaternary semiconductor crystal can be grown on an InP substrate with lattice matching. Liquid-phase epitaxy is the commonly employed method in production, but vapor-phase epitaxies such as MOCVD (metal–organic chemical vapor deposition) and CBE (chemical beam epitaxy), or MOMBE (metal–organic molecular beam epitaxy) have also been studied. For optical communications, a refractive index guiding laser (see Chapter 5) with new longitudinal modes and highly linear current input-to-light output characteristics is most often used. The growth method is classified into buried types and structured substrate types with V- or U-shaped grooves. A typical laser has a threshold of approximately 20 mA, a differential quantum efficiency of 80%, and an output power of 10–20 mW, but 150–300 mW is sometimes reported for experimental devices. PCM at 400 Mbit/sec is now in practical use. High-speed modulation at 1.6 Gbit/sec is now available and multi-gigabit per second or higher modulation rates are being studied experimentally. Devices with high reliability for long-haul transmission system and undersea cables are being developed. Lifetimes of 10^5 to 10^6 h have been measured by testing at high temperature to accelerate the degradation rate.

When a semiconductor laser is modulated at high speeds, longitudinal modes are increased in number and large dispersion occurs, especially at wavelengths around 1.55 μm in the lowest loss band of the normal silica fiber. These modes must be suppressed for high-speed transmission. Therefore, research on dynamic single mode (DSM) lasers, where the number of longitudinal modes does not increase even with high-speed modulation, is progressing at a remarkable pace.

Some industries are producing distributed feedback lasers (DFB) to drive optical transmission lines at 1.6 Gbit/sec. However, even the DFB laser has a dynamic wavelength shift or chirping when modulated at high speed. This is the result of a change in the refractive index through a slight fluctuation of carriers. The value of the frequency fluctuation Δf is written as:

$$\Delta f = \frac{\alpha}{4\pi}\left[\frac{1}{S(t)}\cdot\frac{dS(t)}{dt} + \text{nonlinear term}\right] \tag{2.1}$$

where $S(t)$ is an optical pulse form and α is a linewidth enhancement factor. A structure to reduce chirping and the use of an external modulator have been proposed as possible approaches to alleviate this problem. In external modulators, chirping also occurs because of a phase modulation. The chirp, Δf, is expressed by an equation similar to that above:

$$\Delta f = \frac{\alpha}{4\pi}\cdot\frac{1}{S}\cdot\frac{dS}{dt} \tag{2.2}$$

This equation differs from that for a laser, however, since the chirping in the external modulator does not interact with relaxation oscillations. Optimization of the modulation waveform leads to a reduction of chirping.[6]

2.2.3. Lasers as Light Sources in Communication Measurement

In order to check for connection faults or fracture points in optical fibers, optical time domain reflectometry (OTDR) is used. When a short optical pulse is injected into the input end of an optical fiber and propagated, the light may be scattered backward if some small refractive index fluctuations exist. In the absence of faults or fractures, this is mainly caused by Rayleigh scattering. Since this backscattered light is from propagating pulses, the observed Rayleigh intensity decreases monotonically with time. If there is a splice, connector, or undesired fault point at some distant location in the fiber, a strong scattering occurs and a signal appears as a spike on the time trace because of the local fluctuation of the refractive index.

According to the time lag, the fault location can be found by determining where the reflection occurs in the fiber. The OTDR is essential for installing fiber cables. A high-output laser is required as a light source. Using the low-loss wavelength band at 1.55 μm enables detection of fracture points as far as 130 km away. GaInAsP/InP semiconductor lasers with p-type substrates have been developed for this purpose.

2.2.4. Optical Amplifiers

If we can directly amplify the optical wave, then the output power from the transmitter can be dramatically increased, and the design choices for lightwave communication systems will be widened. As described in Chapter 4, it seems relatively straightforward to make an optical amplifier, when we remember the fact that the induced emission is basically an amplifying process. Actually, a multiple series of optical amplifiers has been utilized in large-scale laser systems such as CO_2 or glass lasers for experiments in laser fusion. The reason the optical amplifier has not been used in the area of lightwave communications may be that the output power of laser diodes can be as high as 10 mW, and that is enough for ordinary uses. However, the driving forces for optical amplifier research have increased since 1988 when the lightwave systems began to require more versatile sources and demand very high performances.

There are several schemes for optical amplification, three being the semiconductor laser amplifier, the optical fiber amplifier, and the stimulated Raman amplifier. The semiconductor laser amplifier is basically a semiconductor laser, but with very small reflection at facets to avoid the laser oscillation. The required reflectivity at the facets should be 0.01% or less for the purpose of minimizing optical feedback and resonance. The light is amplified by simply using the single path gain of the medium as a traveling wave amplifier (TWA). With the GaInAsP/InP amplifier in the wavelength band of 1.55 μm, more than 20 dB of gain can be provided from fiber to fiber. The handling power can be as high as 10 mW. One problem of the semiconductor laser amplifier is the existence of a gain difference between the TE and the TM modes. The gain for the TE mode, whose electric field is parallel to the junction plane, is generally larger than that of the TM mode, whose magnetic field is parallel to the junction plane. This is caused by the difference of confinement factors associated with the TE and TM modes. The light transmitted through the optical fibers is normally depolarized. In that case, TE and TM components are amplified differently and intensity noise may be generated when the polarization state fluctuates. In order to avoid this problem, some special designs of amplifier structure are considered: (1) making the waveguide structure symmetric to equalize the confinement factors for TE and TM modes and (2) introducing a strained quantum well with tensile stress which generates a higher gain for TM modes. A gain difference of less than 0.5 dB has been attained by using these techniques.

The optical fiber amplifier uses the induced emission of erbium (Er) ions which have been doped into the optical fiber piece so that it functions as an amplifier. The Er ions are pumped by a short-wavelength light source, which is also a semiconductor laser. This type of optical amplifier has several advantages, among them being (1) mode-matching is easy since the amplifier is basically composed of the same fiber, (2) there are small differential gains for TE and TM modes, and (3) the temperature dependence is small. The disadvantages include the fact that the gain bandwidth and the center operation wavelength are limited. To get enough gain, more than 40 mW of excitation power is required from the pumping lasers. To match the Er gain resonances, a 1.4-μm GaInAsP/InP laser or a 0.98-μm GaInAs/GaAs strained quantum well laser is utilized.

If some high-performance amplifier is finally realized, the following new features (among others) in lightwave systems will be possible: (1) the output power in distributed systems can be remarkably increased for many customers, (2) optical preamplifiers become possible, and (3) nonregenerate repeaters become available. In 1989, approximately 904-km and 2000-km transmission experiments using more than ten amplifiers were conducted to demonstrate the feasibility of fiber amplifiers.

2.3. LASER DISKS

As is shown in Fig. 2.3, laser light is focused on a disk by using an objective lens with a large numerical aperture (NA) to read out the patterns of

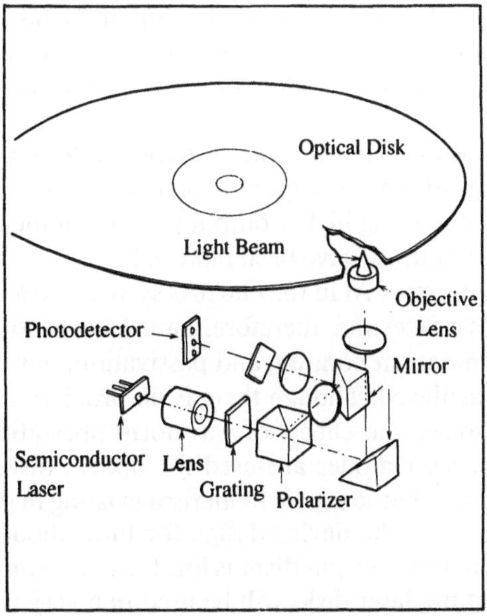

Figure 2.3. Optical disk. [After Y. Mitsuhashi.]

pits through the difference in equivalent refractive indices.[2] This device is known as an optical disk and has been commercially available since 1984. Needless to say, the laser which is used for the optical disk must be a semiconductor laser. The number of compact disks (CDs) manufactured in 1985 amounted to approximately 4 million, causing the production of approximately 5 million laser chips. Monthly production in 1990 reached on the order of several million.

The spot diameter, ΔD, formed by an object lens is given by:

$$\Delta D = 1.22 \cdot \frac{\lambda}{\mathrm{NA}} \tag{2.3}$$

If the wavelength, λ, is shorter, the lens NA can be smaller; therefore, optical design becomes easier. A GaAlAs/GaAs laser operating at 0.78 μm is most often employed and is fabricated with an active layer grown on a grooved substrate.

If the reflected light from an optical disk returns to the semiconductor laser cavity, intensity noise may increase. Also, longitudinal mode-hopping occurs because of temperature change, resulting in an increase of intensity noise. Some measures which are employed to suppress this are, for example, (1) increasing the number of oscillating modes so averaging occurs and (2) superposing a high-frequency signal to reduce the coherence, especially in the case of analog video disks. The relative intensity noise (RIN), which is defined by equation (13.1), is used to express the intensity noise. A value of -140 dB/Hz for RIN is considered to be a requirement for optical disk systems.

An output power of a few milliwatts is enough for reading out, whereas a power exceeding 40 mW is required for writing into a disk. Therefore, research aimed at obtaining higher output power is under way, and powers exceeding 200 mW or more have been reported.

The chemically active Al in the GaAlAs system must be protected from oxidation at the laser facet and, therefore, some facet passivation is required. Through improvements in bonding and passivation, devices with high reliability have been achieved. Though there is no problem for ordinary applications of laser diodes, the effects of waveform aberration become much more severe when laser diodes are used for optical disks. The transverse phase front is distorted because of astigmatism existing in gain-guided lasers, as shown in Table 2.1. The device design for these diode lasers should be carefully made so that the astigmatism is limited to 2–3 μm or less.

In the near future, laser disks will be used in a very wide variety of application areas.[3] For example, CD-ROMs (fixed memories recorded on an optical disk) are being put into practical use. A disk of more than 500-Mbyte capacity, which is equivalent to the memory stored on a standard CD, becomes possible. This is equivalent to 500 floppy disks. As rewritable optical disks are becoming available, additional very interesting applications will become possible, e.g., storage of documents such as books, pictures, and catalogs, large-scale computer memory, replacing magnetic tapes.

Digital color images for television or video sequences require much larger capacity. The current technology in 1990 enables one to record and play back images for up to 10 min by using two LP record-sized digital disks, each with a capacity of approximately 10 Gbyte. From this example, it is clear that extremely large capacity memories are required to digitally record moving color images. Although optical disks have a total capacity relatively smaller than that of magnetic tapes, their capability of quick playback and random access is a unique advantage and will be utilized for broadcasting and picture communications.

2.4. LIGHTWAVE SENSING

Lightwave sensing[4] using the coherent characteristics of lasers is of increasing importance and has led to the development of various new applications such as laser Doppler velocimeters[5] and optical gyroscopes.[6] There are some problems, however, in using conventional semiconductor laser diodes as optical sources in lightwave sensing. These problems are mostly caused by the device structure and strongly affect the laser performance in a measurement system. It is desirable to use practical devices which are mainly fabricated for optical disks and optical fiber systems without disturbing their optical characteristics. By carefully applying these lasers, we can improve their performance to some extent (especially toward high sensitivity and low drift) in order to meet the requirements for lightwave sensing. Some of the problems and remedies associated with conventional semiconductor lasers are outlined in Table 2.2.

2.4.1. Light Output

A conventional laser diode can emit a continuous output of 1–10 mW at room temperature, but this is still low for some sensing applications. In order to take full advantage of the laser output, we must use low-loss optical devices and low-loss optical transmission media. At the research level, progress is being made in getting higher output, with >200 mW now available

Table 2.2. Performance Characteristics of LDs, Problems with Light Sources and Their Remedies

Performance characteristics	Problems	Remedies	Improved device design for sensor systems
1. Output = 1–5 mW power	Low sensitivity	Use a low-loss optical device for an optical transmission system	Aiming at large output (>100 mW) and arraying
2. Emission angle of bundle of light = 15–30°, asymmetry	Small focusing efficiency, disorder of beam spot and wave front	Use small-sized AR-coated collimating lenses	Surface-emitting laser
3. Facet reflectivity 30%	Unstable oscillation by reflection	Use a temperature-controlled isolator, or replace with a multimode laser or SLD	Reflectivity control
4. Power variation due to temperature −50 μW/K	Increase of drifts	Temperature control	Monolithic APC
5. Wavelength variation by temperature +1 nm/K	Decrease of coherency	Temperature control	Monolithic AFC
6. Mode hopping	Low sensitivity, difficulty in wave tuning	Temperature control	DSM laser
7. Wavelength variation ± 20 nm	Difficulty in wave tuning	Temperature control	DBR laser
8. Oscillator spectral width 10 to 100 MHz	Decrease of coherency	Frequency control	External resonator-type
9. Antielectric surge ≦ 500 V	Short lifetime (actual working hours: 10^3)	Assemble with care Low-noise power supply is needed	Crystal quality Strengthen junction

from a single laser chip, and an output of several watts expected using an array structure.

2.4.2. Radiation Angle of Light Beam

Emitted light from a laser diode which has a (5×1)-μm elliptical near-field beam pattern has an asymmetric far-field pattern with a half-angle divergence of between 15 and 30° caused by diffraction. This results in low coupling efficiency to optical fibers and difficulty in generation of collimated beams, leading to nonuniform intensity and phase distributions. This frequently yields poor isolation if an optical isolator is employed in the beam path. The collimating microlens must have an antireflection coating to avoid feedback noise, and, even then, feedback noise occurs because of the reflection from the lens when the coating is not perfect. A solution to this may be the surface-emitting laser diode, which has a small angle of radiation. This device will be discussed later.

2.4.3. Instability of Laser Operation Caused by Optical Feedback

The facet reflectivity of laser diodes, which is determined by the value of the refractive index of the semiconductor material, is approximately 30%. Light fed back into the laser diode from other optical devices can easily enter the cavity because of this low reflectivity. The performance of a lightwave sensing system will deteriorate as a result of the instability of the laser operation (feedback noise) which is induced by the light fed back into the cavity. Since this instability occurs even when the level of feedback light is less than 0.01%, feedback must be strongly suppressed.

An optical isolator using the Faraday effect is generally used to prevent reflected light from coupling back to the laser oscillator. Usually, lead glass for the wavelength range of 0.7–0.8 μm and YIG for the wavelength range of 1.3–1.6 μm are used as Faraday rotators. These can produce an isolation of about −30 dB. The Faraday rotation angle for the YIG has a temperature dependence of about −0.26 K/cm and +2.0 K/cm for wavelength ranges of 1.15 and 1.5 μm, respectively. Therefore, either some temperature control is required, or a new optical isolator material must be developed which is insensitive to temperature. Multimode lasers for video disks and superluminescent light-emitting diodes for nonresonant fiber gyros are used to average out the influence of the feedback light. Sometimes, the reflectivity of the facet coating is increased when the device is made so as to provide some additional isolation. This technology will be used more frequently in the future.

2.4.4. Increase of Intensity Noise Caused by Mode Hopping

Mode hopping occurs from one specific longitudinal mode to another mode in the Fabry–Perot laser resonator. The frequency of this hopping is in the range of $0.1-10^7$ per sec, and decreases with increasing bias current. Moreover, this kind of laser has a hysteresis in its current-to-laser-intensity characteristic. This phenomenon can be thought to be caused by spontaneous emission which excites all longitudinal modes, and some specific longitudinal mode is enhanced over the others at a particular instant in time. Because of mode hopping, intensity noise increases and the sensitivity of lightwave sensing systems deteriorates and it also becomes difficult to continuously tune the wavelength. By controlling the temperature and the current, mode hopping can be suppressed. The dynamic single mode laser (DFB or DBR type) is potentially a laser with no mode hopping. For lightwave sensing systems, it is desirable to have lasers which operate in the range of 1 μm or less.

2.4.5. Variation of Power Related to Temperature Change

If the ambient temperature of a laser device changes, its threshold current and quantum efficiency vary, and, accordingly, the output power changes. The variation is about $-50\ \mu W/K$. This kind of change is seen as a drift by lightwave sensors, and some automatic power controller (APC), with the help of an external electric circuit, is required to suppress the temperature change. It is desirable to have a controller in an electronic device which is monolithically integrated with the semiconductor laser. As a first step toward this goal, a laser with an integrated monitoring detector is currently being developed.

2.4.6. Variation of Wavelength Related to Temperature Change

The wavelength of conventional Fabry–Perot laser diodes exhibits a shift caused by the temperature change of about $+5$ Å/K for both the GaAs and GaInAsP systems. Changes in wavelength decrease the temporal coherence of the light and usually cause a reduction in the sensitivity and drift in lightwave sensing applications. In order to compensate for this change, temperature control as well as current control are employed. Since the output power is also varied when the current changes, special care must be paid to the total system design. As before, it is desirable to integrate the automatic frequency controller into a monolithic structure with the semiconductor laser.

Research on developing a laser with an integrated frequency monitor and wavelength tuner is under way. One important task associated with this research is to develop a frequency standard which is not itself subject to temperature drift and to which the laser can be frequency locked.

2.4.7. Reproducibility of Wavelength

The variation of lasing wavelengths among devices is about ±50 Å in the case of GaAs lasers. Some of this variation can be accommodated by wavelength tuning by changing the injection current, but it is rather difficult to tune to a particular wavelength because of longitudinal mode hopping. For example, if we want to adjust the laser wavelength to the center of a particular absorption line of a gas, a precision of about 0.01 cm^{-1} of wavenumber is needed. If we select a suitable laser from commercially available devices, only 10–25% can be operated at the specified wavelength.

In order to widen the wavelength tuning range, the applicability of using an external grating mirror has been demonstrated. There are, however, some difficulties in using this method in a practical system because of the instability caused by reflected light and the reliability of the total system. It has also been reported that some temporal variation of wavelength occurs, particularly in short-wavelength GaAlAs/GaAs lasers. The variation of thermal resistance after aging and the thermal effects relating to the nonradiated free combination of carriers in the laser cavity are possible causes of this effect. A quantitative investigation of the origin of this effect is important so that the result can be taken into consideration at the time of device design.

2.4.8. Linewidth

One of the most important criteria for a coherent light source is the linewidth. The reported values of linewidth typically range from 10 to 100 MHz. On the other hand, the requirement for coherent communication is 1 MHz, and that for resonant-type fiber gyroscopes or Doppler velocimeters is 0.1 MHz or less. To fulfill these requirements, some artificial narrowing of the linewidth is necessary. Since the linewidth directly affects the sensitivity of coherent sensors, this will be one of the key technologies to be developed in this field.

For the purpose of linewidth narrowing, it is essential to increase the Q value of the laser resonator. By employing a high-reflectivity external mirror, a linewidth of 1 kHz has been demonstrated. An optical fiber is sometimes used to elongate the cavity, and a 30-kHz linewidth has been reported using this technique. These lasers still have problems with instability caused by optical feedback which brings about widening of the linewidth, large optical volume of the system, a deterioration of modulation characteristics by current injection, and so on. In order to overcome these problems, a novel method to suppress FM noise by negative feedback of the injection current in wide frequency ranges has been developed. By careful experimentation using this technique, <1 Hz of linewidth has been reported. In principle, a linewidth of about 1 mHz to 1 Hz is possible. That is narrower than the expected value from the well-known Schawlow–Townes relation.

2.4.9. Resistivity against Electrical Surge and Lifetime

Laser diodes can be expected to survive surge voltages of about 500 V. But high voltage sometimes damages the device rapidly, so it is preferable to employ a stable current source without any current noise.[6,7]

By including the wavelength variation, the actual lifetime of laser diodes is thought to be about 10^3–10^4 hr. To raise the reliability of devices for stable lightwave sensors, for example, the quality of semiconductor crystals and careful formation of a *pn* junction must be considered.

2.5. ELECTRO-OPTIC EQUIPMENT

In the optoelectronics field, much equipment utilizing lasers is being developed.[8] Of course, optical fiber communications, optical disk memories, and lightwave sensors are major areas. Here we discuss other applications such as electro-optic equipment.

Bar code readers are frequently seen, for example, at checkout counters in supermarkets. By scanning a laser beam along the bar code, a code number is read-in which identifies the product and its price. Presently, He–Ne lasers are employed in this application, but in the near future, semiconductor lasers emitting at 0.63 μm will become available and will replace the He–Ne lasers.

A laser beam printer is one which uses a scanned and focused laser beam with intensity modulation to write the characters to be printed. Its printing speed is very high and it is quiet relative to impact-type printers. A high-power semiconductor laser is being introduced for this application. In present machines, a polygonal mirror is utilized for mechanically scanning the laser beam. This mechanical mechanism limits the printing speed, so an electro-optics scanner is desirable. Laser beam steering is one of the most difficult problems in the laser application field. Extensive work is expected in this area.

2.6. MEDICAL APPLICATIONS

Lasers have been extensively developed for medical applications.[9] They can achieve high energy density and can be tightly focused for surgical procedures, and their spatial and spectral qualities can be used for chemical and biochemical reactions.

One of the earliest and probably the best known medical application is the laser coagulation system used in ophthalmology. Formerly, operations on retinas were performed using a fine needle from the far side of the eyeball. Alternatively, the coagulation of bleeding vessels was achieved using a xenon lamp. Now, however, if one uses a laser light in the operation, the focused spot on the retina can be made very small. Thus, a safer and simpler operation becomes possible except for special cases. In this application an Ar-ion laser is often used with approximately 2 W of output power. In surgical procedures, CO_2, Ar, and YAG lasers are used as a laser "knife." A laser light with high energy density can cut tissue simultaneously performing dehydration, shrinkage, carbonization, and evaporation. Since the laser is also effective in stopping the bleeding, operation times can be shortened which means safer and more localized procedures. The laser does not touch the tissue, so a high degree of sterility is maintained, which is particularly important for the treatment of burns. Lasers can be effective in delicate procedures such as brain surgery.

In internal medicine, the laser light can be guided through a laparoscope to vaporize endometriosis and accomplish other gynecological procedures or passed through an optical fiber into the bladder to destroy kidney stones. A fiber optics utilizing a bundle of fibers is a concept which was invented and developed in Japan. Research on cancer identification and treatment using specific wavelengths of light is currently being performed. Basic research is also progressing towards wider applications, such as laser irradiation of cells, measurement of blood speed, and the observation of eye shapes and brains by the use of three-dimensional holography.

2.7. ENERGY DEVELOPMENT

Laser fusion projects are being developed in many countries as a part of national efforts toward the goal of realizing fusion.[9] The lasers achieve high-density, short-time plasma confinement by an implosion driven by high-energy light illumination. Extremely high temperature and high density are generated by irradiating a spherical target pellet with an intense laser beam whose power is on the order of terawatts ($1\ TW = 10^{12}\ W$). The outer part of the small spherical pellet is explosively ablated, and a high-temperature plasma is generated in the center as a result of the spherical compression wave. Achieving a critical temperature, and sufficient confinement time and density to obtain fusion is, of course, the object of this project.

Examples of other industrial applications include laser manufacturing and laser-assisted processes where high-power CO_2, YAG, and excimer lasers are employed in heat treatment, welding, and drilling. On the other hand, laser cutting of iron plates and clothes by computer control is becoming a very effective method. Microfabrication by using the good focusing ability of laser beams is an important application for scribing, marking, and frequency tuning of crystal oscillators, etc. Various laser-assisted processes are now being developed mostly using ultraviolet lasers to enhance such things as chemical reactions and chemical separation. These include such applications as laser etching, laser hardening of plastics, laser chemical vapor deposition (CVD), and laser isotope separation.

2.8. LASER DISPLAY

It is very interesting to apply the laser to the area of display in holographic arts where the demands are somewhat different from those of a technical field, but take advantage of coherence, narrow beam behavior, and high radiance. Dynamically moving images synchronized with background music can be produced by laser beam scanning and projected on walls, ceilings, or clouds. For this purpose, Ar and Kr high-power ion lasers are usually used. Holographic art constitutes the three-dimensional regeneration of images by using interference of laser light. Holograms are recorded by interfering the coherence signals scattered from the objects and a uniform reference beam at the film plane.[10,11] Application of holograms to art and to identification emblems on credit cards are good examples of this technology.

PROBLEMS

2.1. What are important characteristics of lasers regarding their application to various engineering fields?
2.2. What performance is needed from lasers for optical fiber communications?
2.3. What is the frequency chirping of semiconductor lasers? Explain its physical origins.
2.4. Explain how frequency chirping arises in an external modulator.
2.5. How are lasers applied as measurement tools in optical fiber communications?
2.6. Explain the principle of an optical disk.

2.7. What is lightwave sensing? What are the points which should be considered in designing lasers for lightwave sensing?
2.8. How is a laser applied in laser printers?
2.9. How is a laser utilized in the medical field?
2.10. Explain the principle of a laser fusion.
2.11. Consider how the laser is applied in photo-assisted processes.
2.12. What is a hologram? How is it applied in engineering and other fields?

REFERENCES

1. For example, Y. Suematsu and K. Iga, *Introduction to Optical Fiber Communication,* Ohmsha, Tokyo (1976); 2nd ed. (1983); 3rd ed. (1992). English version for 1st ed., from Wiley, New York (1978).
2. For example, Y. Mitsuhashi, in *Optical Disk Systems,* Asakura, Tokyo (1989).
3. G. Bouwhuis, J. Bratt, A. Huijser, J. Pasman, G. van Rosmalen, and K. S. Immink, *Principle of Optical Disc System,* Adam Hilger (1985).
4. T. Yajima, K. Shimoda, S. Namba, and H. Inaba, eds., *Laser Handbook,* 2nd ed., Asakura, Tokyo (1989).
5. J. Knuhtsen, E. Olldag, and P. Buchhave, *J. Phys. E* **15,** 1188 (1982).
6. M. Ohtsu and K. Iga, Oyo Butsuri, **54,** 747 (1985).
7. K. Iga and M. Ohtsu, *Fall Meet. Jpn. Soc. Appl. Phys.* 13p-Q-6 (1984).
8. K. Iga, in *Optoelectronics,* Asakura, Tokyo (1984).
9. *Laser Handbook,* The Laser Society of Japan (1982).
10. D. Gabor, *Nature* **166,** 777 (1948).
11. E. N. Leith and J. Upatnieks, *J. Opt. Soc. Am.* **52,** 1123 (1962).

THREE

GAS AND LIQUID LASERS

This chapter describes gas-phase lasers which utilize energy levels of gaseous atoms, ions, and molecules, as well as liquid lasers which use dye molecules dissolved in a liquid. Gaseous molecules exhibit a narrow gain width because they are relatively isolated from the environment, while liquid lasers feature wide gain widths which lead to broad wavelength tuning capabilities. These lasers are generally pumped by gas discharges or by optical excitation.

3.1. GAS LASERS

3.1.1. Helium–Neon Laser

The He–Ne laser is one of the most common gas lasers and was the first gas laser to be realized.[1] Because the individual gas atoms only collide occasionally with other atoms or molecules in the environment, they are not perturbed as much by the surroundings as would be the case in the liquid or solid state. As a consequence, the gain bandwidth is very small. For gas lasers, in general, the oscillation wavelength depends on the kind of atom or molecule which is generating the gain, and relatively high frequency stability can be obtained because of the small change of the gain center frequency as a function of temperature. In an ordinary gas laser, the natural wavelength stability is approximately one part in 10^6 to one part in 10^7. Stabilities of around one part in 10^{12} to one part in 10^{14} are obtainable through automatic stabilization. We note that the frequency stability is the ratio of the frequency fluctuation, Δf, to the center frequency f_0. Therefore, a stability of 10^{-14} for

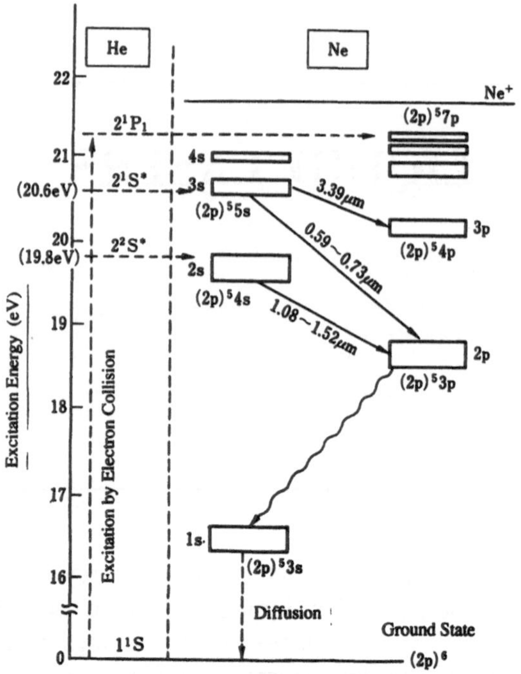

Figure 3.1. Energy levels of He and Ne atoms. [After A. Javan, W. R. Bennett, Jr., and D. R. Herriott, *Phys. Rev. Lett.* **6**, 106 (1961).]

a visible laser such as the He–Ne laser corresponds to an absolute frequency stability on the order of 5 Hz. Therefore, gas lasers are particularly well suited for applications which require high frequency stability.

The He–Ne laser is one of the most commonly manufactured lasers, and the red line at 6328 Å is widely used. There are, however, other laser emission lines at 1.15 μm, 3.39 μm, and at other wavelengths in the near infrared. Figure 3.1 shows the energy level of the Ne atom. Helium atoms are excited to high-lying energy states (2^3S^*, 2^1S^*, 2^1P^1, etc.) by an electric discharge. The excitation energy is then transferred by collision to the Ne atom which subsequently lases, as shown in the figure. He–Ne lasers of 1 mW to 100 mW are commercially available and have a high reliability for a wide range of applications. Figure 3.2a shows the experimental setup for a pure Ne laser that was studied by the author. In this case, the Ne atoms are directly excited by a gas discharge. The output power is then recorded as the cavity length is changed by a voltage applied to a piezoelectric transducer (PZT). This tunes the cavity mode through the gain curve of the neon, and, as is theoretically discussed in Chapter 11, a Lamb dip at the center of the gain curve can be observed. (The output wavelength is 1.15 μm.)

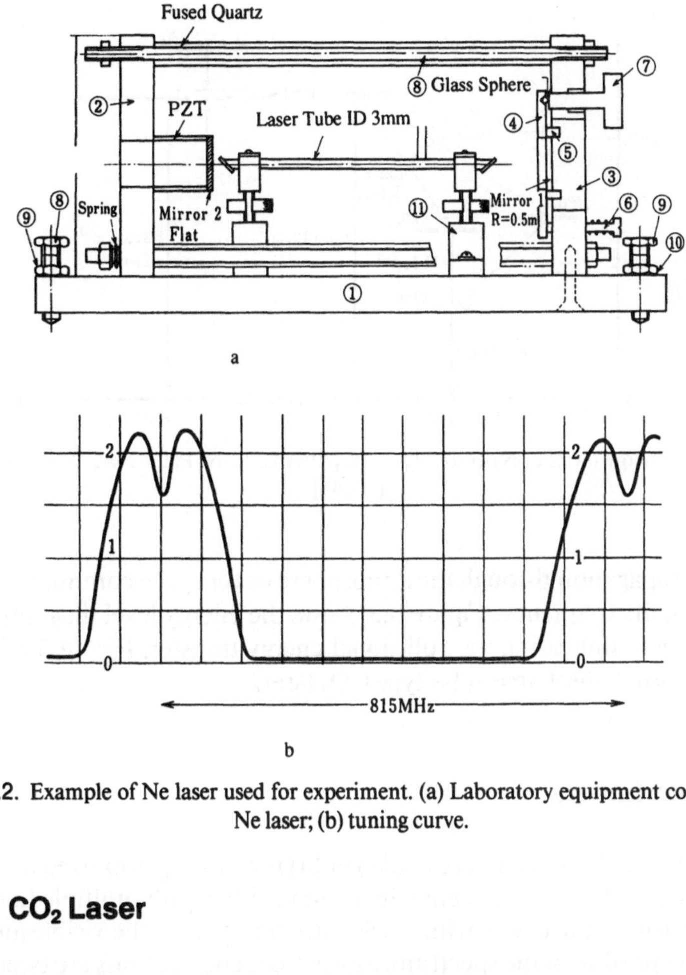

Figure 3.2. Example of Ne laser used for experiment. (a) Laboratory equipment comprising a Ne laser; (b) tuning curve.

3.1.2. CO_2 Laser

The CO_2 laser[2] oscillates mainly in the infrared region at 10.6 μm and is one of the high-power lasers. Excitation is usually done by using an electric discharge to vibrationally excite nitrogen molecules. These then transfer their energy by collision to the CO_2 molecule and isometric stretch mode. This then lases to the symmetric stretch mode, generating numerous laser lines arising from the various rotational levels. Some 10 W of continuous wave (cw) output can be obtained from the simple discharge tube, and more than 20 kW can be obtained using more sophisticated designs. CO_2 lasers ranging from several watts to several tens of watts are utilized for industrial applications. The higher-power CO_2 lasers are used for cutting and welding. The CO_2 laser may be *Q*-switched, generating peak output of approximately 1 MW. The wavelength of 10.6 μm corresponds to a window in the atmosphere, and the CO_2 laser was originally studied as a source for communica-

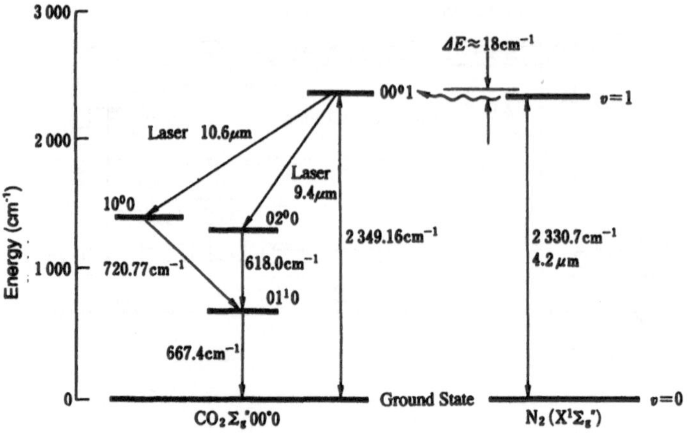

Figure 3.3. Energy levels of N_2 and CO_2 lasers. [After C. K. N. Patel, *Phys. Rev. Lett.* **12**, 588 (1964).]

tion by propagation through the atmosphere or for space communication by stabilizing the frequency. Figure 3.3 shows the energy level diagram of CO_2 and nitrogen, indicating the collisional energy transfer. Figure 3.4 is a diagram of a small discharge-tube-type CO_2 laser.

3.1.3. Ion Lasers

Ion lasers[3,4] use either argon (Ar) or krypton (Kr) gas ionized by an electric discharge. These lasers generally operate either with multiple lines or by using a prism in the cavity with single lines throughout the visible and near-ultraviolet portion of the spectrum. Single frequency outputs are typically on the order of 1 or 2 W, and multiple line output can be up to 10 or 20 W. Since these lasers have many lines throughout the visible, they are often used

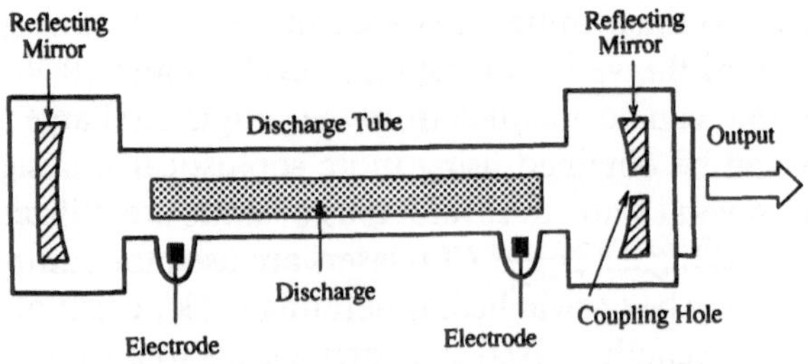

Figure 3.4. Small-sized equipment of a CO_2 laser.

Table 3.1. Oscillating Lines of Ar Ions

4545 (Å)	Weak
4579	Ordinary
4658	Ordinary
4765	Ordinary
4880	Strong
4965	Ordinary
5145	Strong

as light sources for displays. Their high power makes them desirable as devices which are used for observing and applying interference phenomena, such as laser Doppler velocimetry, and their high cw output at short wavelengths makes them useful, for example, for the formation of original optical disks. Tables 3.1 and 3.2 show oscillating lines of argon ions and krypton ions, respectively. Figure 3.5 shows a schematic view of an argon-ion laser tube.

3.1.4. Helium–Cadmium Laser

The He–Cd laser[5,6] uses an electric discharge in a metallic ion vapor (cadmium) to generate blue light at 4416 Å, and ultraviolet light at 3300 Å. This is one of the few cw short-wavelength lasers and its output of several tens of milliwatts is utilized for a variety of applications including the formation of diffraction gratings through a holographic interference method.

3.1.5. Nitrogen Laser

The nitrogen (N_2) laser[7,8] can only operate as a pulse source and generate a high-power output in the ultraviolet at 3370 Å, with pulse widths on the

Table 3.2. Oscillating Lines of Kr Ions

4680 (Å)
4762
4767
4825
5208
5309
5682
6471
6764

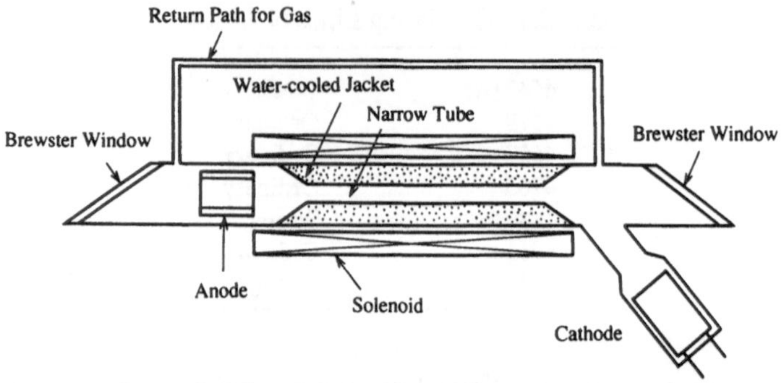

Figure 3.5. Components of an argon-ion laser. (After Refs. 3 and 4)

order of 2 nsec. Since average output power in excess of 100 mW can be achieved, the N_2 laser is often used as an optical pumping source for dye lasers. By itself, however, the N_2 laser has poor coherence, so its applications are limited.

3.2. EXCIMER LASERS

Excimer is an abbreviation for "excited dimer." Atoms that in the normal ground state do not bond together may, in the excited state, form an excimer molecule.[9] Lasing then occurs from instantaneous induced emission from this excited dimer state back down to the ground state, whereupon the molecule immediately dissociates. Electron beam excitation or electric discharges are generally used for pumping. Table 3.3 shows the wavelengths of various excimer lasers.

The excimer lasers are particularly useful since they emit ultraviolet laser beams with wavelengths shorter than 2000 Å, i.e., photon energies from 3 to ~10 eV. These lasers are being utilized for chemical processing and etching such as photochemical reactions, for the formation of films, hardening of plastics, and nonintrusive diagnostics. Lithography, photoengraving, and localized germ killing processes are other applications which take advantage of the short wavelength and relatively good mode of these lasers.

Table 3.3. Oscillation Wavelength of Excimer Lasers

	F_2	ArF	KrF	XeCl	XeF
Gas wavelength (Å)	1570	1930	2490	3080	3500

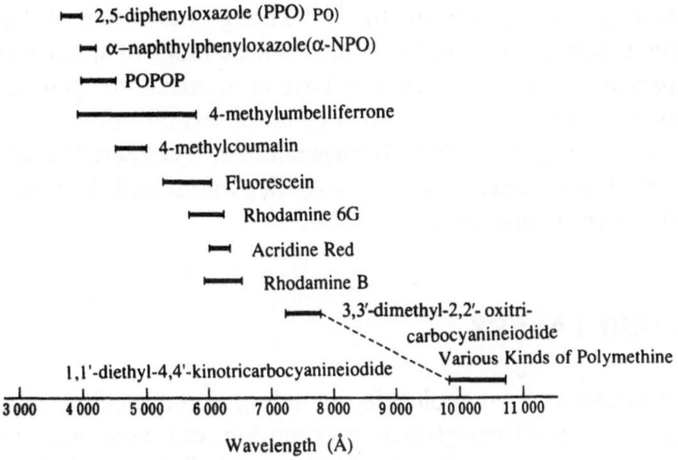

Figure 3.6. Oscillation region of a dye laser.

3.3. LIQUID LASERS

A typical example of a liquid dye laser is that which uses rhodamine 6G dye mixed in alcohol and excited by either an argon-ion laser, a frequency-doubled Nd:YAG laser, a nitrogen laser, an excimer laser, or a strong short-pulse flash lamp to obtain lasing in the red portion of the spectrum. The dye has an emission band of more than 100 Å, so its output can be tuned by changing the cavity resonance frequency. By selecting from various dyes such as those shown in Fig. 3.6, the entire wavelength range visible to infrared can be covered. Figure 3.7 shows the absorption and fluorescence spectra

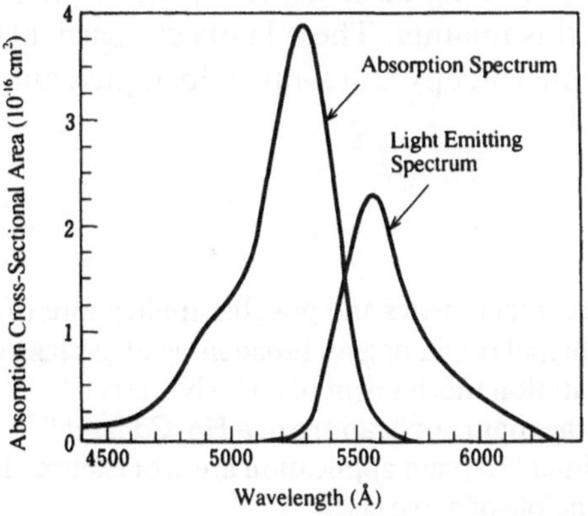

Figure 3.7. Absorption and fluorescence spectra of rhodamine 6G.

of a typical dye, i.e., rhodamine 6G.[10,11] Though the output of the dye laser is typically of low power, dye lasers have a wide range of applications. Their broad spectrum makes them particularly appropriate for generating short pulses and using mode-locking and pulse compression techniques, dye lasers as short as 5 fsec (10^{-15} sec) have been generated. Dye lasers may also be used for high-resolution spectroscopy as a result of their wide tuning range and potentially narrow linewidth.

3.4. OTHER LASERS

In virtually any gas molecule, including water, alcohol, etc., induced emission is possible through strong excitation and laser operation can be obtained. There are many lasers which use metallic vapors[12]; however, the one which uses copper vapor features a visible light output in three primary colors, i.e., red, green, and blue. By taking advantage of a high output power in the visible region, this laser may be applied in some isotope separation experiments.

Lasers using chemical reactions, particularly deuterium plus fluorine, generating deuterium fluoride, have been developed for high-power applications and military uses. Free electron lasers use a high-energy, linear accelerator to generate a high-speed electron beam which is then modulated in a Wigler field so that the electrons emit coherent strong radiation in the forward direction. This generates a high-energy laser beam which is frequency tunable.

Recently, x-ray laser beams have been generated using high-energy, recombining plasmas. These lasers are made by focusing very high power beams from CO_2 lasers or solid-state lasers onto small targets to generate highly ionized, recombining plasmas. Laser beams at 182 and 207 Å have been generated in this manner. These lasers are particularly exciting for instantaneous x-ray microscopy and x-ray holography, among other things.

PROBLEMS

3.1. Explain typical characteristics and possible applications of gas lasers.
3.2. What is the principal origin of gain broadening of gas lasers?
3.3. Explain the excitation mechanism of a He–Ne laser.
3.4. What is one of the main applications of a He–Cd laser?
3.5. What is an excimer? Explain application areas of excimer lasers.
3.6. Explain the principle of a dye laser.

REFERENCES

1. A. Javan, W. R. Bennett, Jr., and D. R. Herriott, *Phys. Rev. Lett.* **6,** 106 (1961).
2. C. K. N. Patel, *Phys. Rev. Lett.* **12,** 588 (1964).
3. W. B. Bridges and A. N. Chester, *Appl. Opt.* **4,** 573 (1965).
4. E. F. Labuda, E. I. Gordon, and R. C. Miller, *IEEE J. Quantum Electron.* **QE-1,** 273 (1965).
5. W. T. Silfvast, *Appl. Phys. Lett.* **27,** 1489 (1971).
6. M. Mori *et al., J. Appl. Phys.* **48,** 2226 (1977).
7. D. C. Cartwright, *Phys. Rev.* **A2,** 1331 (1970).
8. H. G. Heard, *Nature* **200,** 667 (1963).
9. N. Basov, V. A. Danilychev, and Y. M. Papov, *Sov. J. Quantum Electron.* **1,** 18 (1971).
10. F. P. Schafer, ed., *Dye Lasers,* Springer-Verlag, Berlin (1973).
11. M. Maeda, *Spectroscopy* **29,** 279 (1980).
12. L. Csillag, M. Janossy, K. Rozsa, and T. Salamon, *Phys. Lett.* **50A,** 13 (1974).

FOUR

SOLID-STATE LASERS

Several kinds of solid-state lasers and their characteristics are described in this chapter. Solid-state lasers contain ions that undergo induced emission within host solid crystals and glasses, and therefore, the emission wavelength of the laser generally depends on the type of doped ions. The representative ion groups are transition metal ions (e.g., Cr^{3+}, Ti^{3+}, V^{2+}, Co^{2+}) and rare earth ions (e.g., Nd^{3+}, Er^{3+}, Pr^{3+}). Since crystals are not conductive, optical pumping is used for excitation. The ruby laser was the first of many lasers to be demonstrated and it serves as an example of what is known as a three-level laser. The YAG lasers have been developed, which emit comparatively large output powers and are widely utilized for research, industrial processing, and medicine. More recently, Ti-doped sapphire lasers were developed for widely tunable lasers replacing dye lasers.

4.1. RUBY LASERS

The ruby laser was first demonstrated by Maiman in 1960.[1] The ruby crystal consists of Cr^{3+} ions which are doped into an Al_2O_3 crystal and emit red light (6943 Å) by optical pumping. The ruby laser continues to be used as a high-energy pulse laser source. It is possible to obtain peak output powers of 100 MW or more by *Q*-switching the resonator cavity, a method by which the loss of the resonator can be rapidly changed so that the energy can be stored in the crystal and dumped out in one giant pulse.

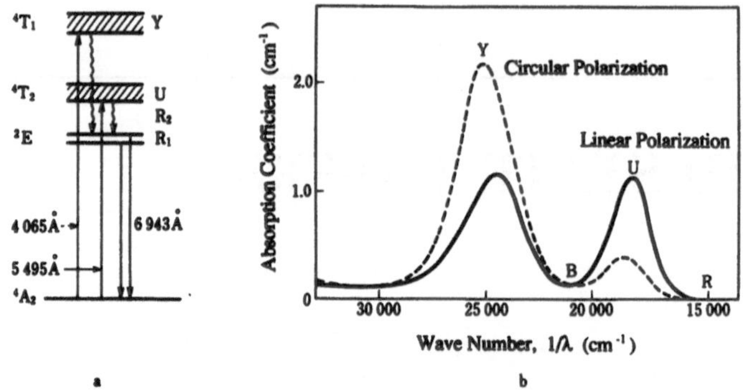

Figure 4.1. (a) Energy levels of Cr^{3+}. (b) Fluorescence spectra of Cr^{3+} [After S. Sugano and Y. Tanabe, *J. Phys. Soc. Jpn.* **13**, 880 (1958).]

Figure 4.1a and b shows the energy levels and fluorescence spectra of Cr^{3+} ions, respectively. The spectral components have been named R, U, B, and Y by Sugano and Tanabe[2] who initially studied the fluorescence of ruby. The Y and U lines are broad absorption bands which absorb light from an extremely strong flash lamp source, and the levels $^4T_1$ and $^4T_2$ are excited. This energy is then transferred by an internal relaxation to the ^2E level so that its population finally exceeds that of the ground state and amplification at the lasing transition, rather than absorption, occurs. In this manner the ruby crystal is activated. The transition from the ^2E state to the $^4A_2$ ground state corresponds to the R line and, if the resonator is formed which reflects light at 6943 Å, laser oscillation is possible.

Figure 4.2a shows an experimental ruby laser which was built by the author around 1962.[3] In Fig. 4.2a, the diameter of one of the reflecting mirrors is reduced in an attempt to control the transverse mode structure of the laser, forming a single transverse mode. This is not unlike current unstable resonator oscillators which are used to couple large amounts of energy out of solid-state laser crystals. This type of approach to transverse mode control was also applied to semiconductor lasers, as will be described later. The relaxation oscillations described previously, are shown in Fig. 4.2a′. Figure 4.2b shows a ruby laser rod with its end cut into a convex mirror,[4] which generates an unstable resonator as discussed in Chapter 8. Figure 4.2b′ shows the relaxation oscillations from this ruby laser.

4.2. YAG LASERS

The Nd:YAG laser (yttrium–aluminum–garnet) and Nd:glass lasers utilize transitions in Nd (Nd^{3+}) ions to generate lasing.[5–7] As shown in Fig. 4.3,

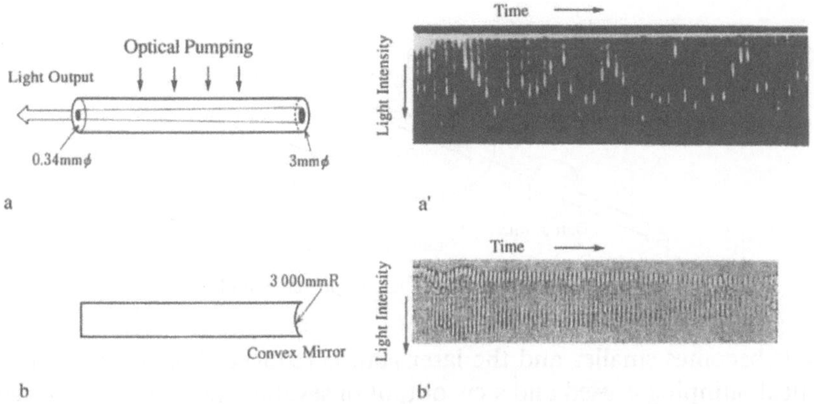

Figure 4.2. Ruby lasers and oscillation pulse waveforms. (a) A ruby laser with a small reflecting mirror; (a′) relaxation oscillation pulse waveforms of the laser in panel a; (b) a ruby laser using a convex mirror (unstable resonator); (b′) relaxation oscillation pulse waveforms of the laser in panel b.

numerous lines are possible, but the highest gain occurs at around 1.064 μm in a Nd:YAG laser. The lower energy level of the lasing transition is significantly far above the ground state that it has a low population. For that reason, population inversions are relatively easy to obtain compared with what is required for ruby lasers where the lower lasing transition is the ground state. Consequently, the ruby laser is known as a three-level laser where the highest level is the one which is pumped, and the lower two levels are the lasing transitions. On the other hand, YAG lasers, along with most other laser systems, are known as four-level lasers since the ground state lies below the lower energy level of the lasing transition. If that lower level is occupied through induced emission, then the population difference between the lasing

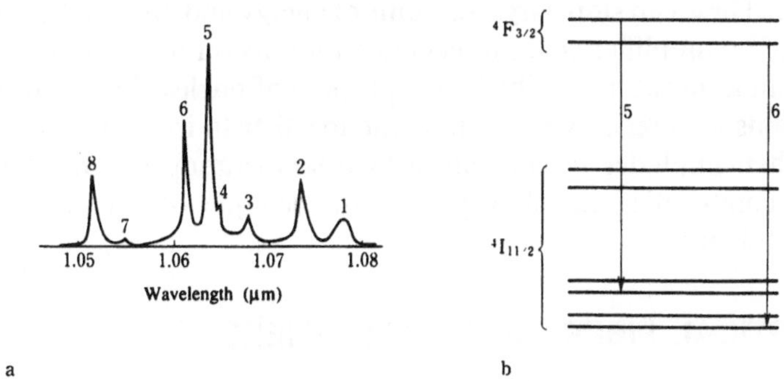

Figure 4.3. (a) Fluorescence spectrum of a YAG (Nd^{3+}) laser and level; (a) Fluorescence spectra of a YAG (Nd^{3+}) laser; (b) Typical spectra and levels [After Ref. 8.].

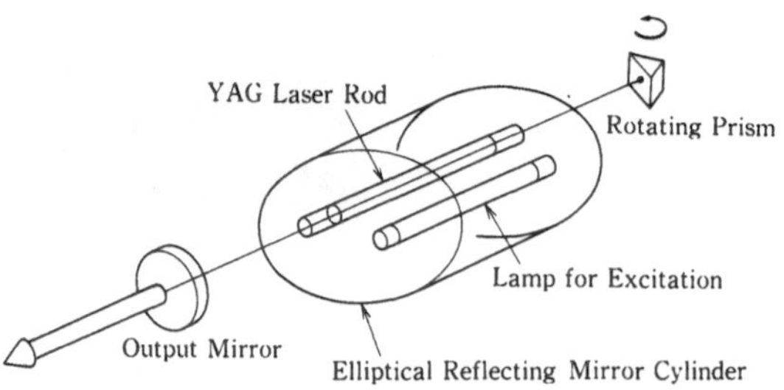

Figure 4.4. Example of a high-power YAG laser.

levels becomes smaller and the laser gain is reduced. For the YAG lasers, optical pumping is used and a cw output of several tens of watts is available at 1.06 μm. These lasers may also be pumped using pulsed flash lamps and Q-switched, generating pulses on the order of 10 nsec long at energies up to approximately 1 J. This corresponds to peak powers on the order of 10^8 W. Much higher energies are possible from Nd:glass lasers because of their larger energy storage capability. These may exceed 1 kJ and are being developed for laser fusion. In many cases, the Nd:YAG and Nd:glass lasers are used in conjunction with second-harmonic-generating crystals which convert the light from 1.06 μm to 0.532 μm with efficiencies as high as 50%. Third-harmonic and even fourth-harmonic generation can be done with high efficiency, making these lasers exceptionally versatile tools. Figure 4.4 shows an example of a high-power YAG laser pumping an oscillator cavity configuration. (This picture is outdated, as I know of no YAG laser using rotating Q-switched prisms that have been built since about 1970.)

4.3. GLASS LASERS

It is possible to fabricate large glass rods and glass disks which contain Nd ions.[8] These can store large amounts of energy and are usually assembled as part of an amplifier chain to generate outputs on the order of 1 kJ with more than terrawatt intensity for the purpose of nuclear fusion. Many such laser beams are precisely timed to come together to irradiate a target pellet which then implodes to generate a high-temperature plasma. Figure 4.5 is a schematic diagram showing the first couple of stages of such an amplifier chain.

4.4. OPTICAL FIBER LASER AMPLIFIERS

Another type of solid-state laser of interest are the fiber laser amplifiers which operate with Er^{3+} or Pr^{3+} ions doped into an optical fiber.

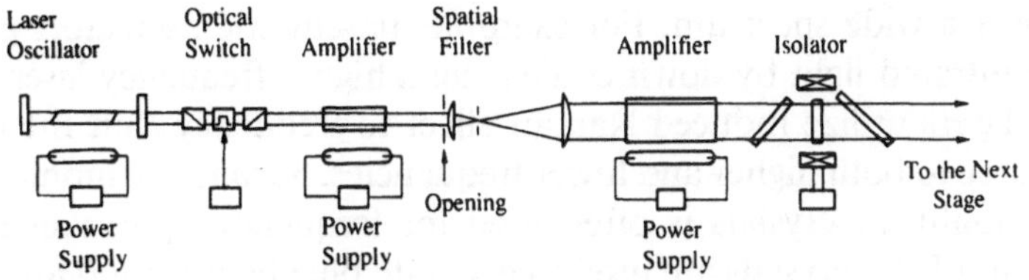

Figure 4.5. Example of a glass laser. [After *Laser Handbook,* p. 227, Laser Society of Japan (1982).]

The laser configuration is shown in Fig. 4.6. The optical gain at 1.55 μm exceeding 30 dB is obtained in Er^{3+}-doped fiber amplifiers. Semiconductor lasers emitting at 0.98 or 1.48 μm are used for pumping. These fiber amplifiers are employed in optical fiber communications.

4.5. OTHER SOLID-STATE LASERS

There are various kinds of other solid-state lasers including some which are miniature devices such as the NdLa phosphorous pentaoxide laser or the LNP ($LiNdP_4O_{12}$) laser which contains Nd^{3+} as one component atom of the crystal and which can provide cw output of several milliwatts using light-emitting diode excitation.[9,10] Other miniature solid-state lasers include tiny YAG lasers which are pumped by small diode laser excitation.

The color center laser is another solid-state laser device which is particularly interesting since it has a tunable output frequency in the vicinity of 1.55 μm, which is the wavelength which corresponds to extremely low loss in silica fibers.[11] This laser operates from transitions in color center defects which are formed in crystals. Its narrow linewidth and tunability make it particularly convenient for spectroscopy. The color center laser has also been used to generate soliton waves which propagate through optical fibers without dispersion.

Various nonlinear optical processes may be used to extend the range of

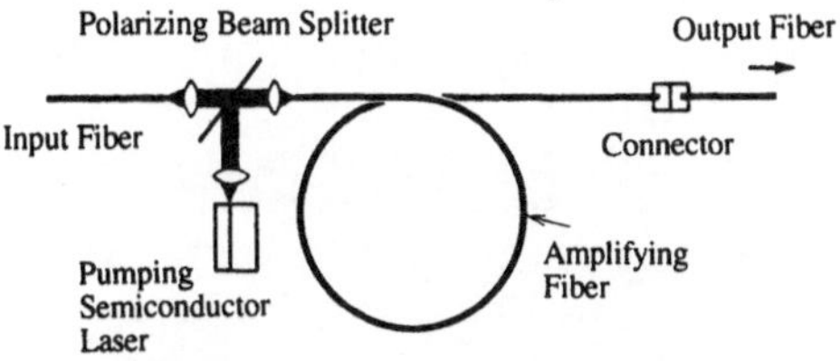

Figure 4.6. Optical fiber amplifier.

lasers over a wide spectrum. For example, parametric oscillators generate tunable infrared light by down-converting a higher-frequency laser source. Raman lasers utilize induced Raman effect to frequency shift the original laser source to both higher and lower frequencies. Second-harmonic generation in nonlinear crystals is often used for frequency up-conversion. $LiNdBO_3$ and KTP crystals are used to generate blue light from 0.8-μm-band semiconductor lasers.

PROBLEMS

4.1. Compare the excitation levels necessary for reaching the threshold of three- and four-level lasers.
4.2. What is a fundamental mechanism of broadening the linewidth of a solid-state laser gain spectrum?
4.3. List possible applications of solid-state lasers.
4.4. Explain the principle of second-harmonic generation.
4.5. What is a soliton laser?

REFERENCES

1. T. H. Maiman, *Nature* **187**, 493 (1960); *Phys. Rev.* **123**, 1145 (1961).
2. S. Sugano and Y. Tanabe, *J. Phys. Soc. Jpn.* **13**, 880 (1958).
3. Y. Suematsu and K. Iga, *Proc. IEEE* **52**, 87 (Jan. 1964).
4. Y. Suematsu and K. Iga, unpublished work (1964).
5. *Laser Handbook*, p. 227, Laser Society of Japan (1982).
6. P. F. Moulton, *CLEO '84 Tech. Dig.*, WA2 (June 1984).
7. J. E. Geusic, H. W. Marcos, and L. G. Van Uitert, *Appl. Phys. Lett.* **4**, 182 (1964).
8. For example, *IEEE J. Quantum Electron.* **QE-17**, No. 8 (1981).
9. H. G. Danielmeyer and H. P. Weber, *IEEE J. Quantum Electron.* **QE-8**, 10 (Oct. 1972).
10. K. Ohtsuka and T. Yamada, *IEEE J. Quantum Electron.* **QE-11**, 845 (Oct. 1975).
11. L. F. Mollenauer and R. H. Stolen, *Opt. Lett.* **1**, 13 (1984).

FIVE

SEMICONDUCTOR LASERS—MATERIALS AND DEVICES

Compact and lightweight semiconductor lasers are available and offer a wide range of oscillation wavelengths throughout the red and infrared portion of the spectrum. These lasers are made by sandwiching the active semiconductor medium between semiconductor material that has a larger band-gap energy and a small index of refraction to confine the optical field. The active medium employs such a double heterostructure which permits the effective confinement of carriers (electrons and holes) and light. With these advances, the semiconductor laser has now developed into a practical industrial laser device. Semiconductor lasers in blue and green wavelength bands are under development.

In this chapter we will outline the development of semiconductor lasers and then discuss the materials used for their fabrication. In addition, we will explain the crystal growth methods that are fundamental to the formation of semiconductor laser devices. This discussion includes such processes as molecular beam epitaxy (MBE), metal–organic chemical vapor deposition (MOCVD), and chemical beam epitaxy (CBE). Finally, we will discuss the fabrication processes of semiconductor lasers. The characteristics and efficiency of semiconductor lasers largely depend on their design and fabrication methods. Therefore, studying these fabrication methods is essential not only for researchers who are involved in device design development, but also for those who wish to put these devices to use.

5.1. OUTLINE OF SEMICONDUCTOR LASERS

5.1.1. Development of Semiconductor Lasers

The semiconductor laser was first realized in 1962 and has now become an essential device for optical communications, optical disk readout, and

Table 5.1. Development of Semiconductor Lasers

Stage	Years	Structure	Concept
I	1962–1969	GaAs homojunction	Direct modulation
II	1970–1976	GaAlAs double heterojunction	Transverse mode control
III	1977–1984	1-μm-band GaInAsP system	DSM
IV	1985–	Short-wavelength region	Variety Quantum well Coherence Control, etc.

other optoelectronic applications. Huge numbers of devices are currently being produced in the industrial sector. We review here the history of semiconductor laser development which can be organized into four stages as shown in Table 5.1.

The first stage began with the birth of the semiconductor injection laser in 1962. Four organizations in the United States (General Electric, Massachusetts Institute of Technology, International Business Machines, and the University of Illinois) developed the homojunction semiconductor laser.[1–4] Several years before this, Braunstein suggested radiative transition in semiconductors,[5] and J. Nishizawa (Japan) applied for a patent on a Te semiconductor oscillator.[6] G. Basov (USSR) proposed the induced emission in semiconductors by carrier injection.[7] There was, in addition, an old suggestion on induced emission by a semiconductor in a notebook of von Neumann.[8]

In the early 1960s, when research on He–Ne, ruby, and other lasers was started, many laboratories and institutions also began research on semiconductor lasers. At that time, J. B. Gunn discovered what is now called the "Gunn effect" in which particular solid-state devices such as gallium arsenide and cadmium sulfide were found to be capable of driving microwave oscillations. As a consequence, studies on III-V semiconductors were started.[9–28] At that time, it was understood that semiconductor lasers must be considered not only as simple diodes, but also as optical waveguides. Expressions and equations leading to a direct modulation limit were also discovered.[2]

The second stage of semiconductor laser development dates to 1970 with the success of continuous room-temperature oscillation achieved by use of double heterostructure devices.[29–41] At that point, we entered the stage of GaAlAs/GaAs semiconductor laser development which had a particularly long duration. This development has been reviewed by I. Hayashi.[41] In 1975, well-balanced optical fiber communication systems using GaAs lasers were realized in combination with silicon photodetectors and optical glass fibers.

In the meantime, the concept of transverse mode control was presented and various structures were proposed to achieve this. Subsequently, a low-loss band in silica optical fibers was found to exist in the region above 1 μm. J. Hsieh (USA), who attended the International Conference on Semiconductor Lasers at Nemunosato in Japan, demonstrated the possibility of a GaInAsP/InP laser with an output wavelength in the 1.1 μm region.

The third stage began in 1977 and featured GaInAsP/InP semiconductor lasers.[42–83] Along with the development of low-loss optical fibers, the reliability of this system was established in a short time, which greatly contributed to the initial realization of optical fiber communications. In 1979, Y. Suematsu (Japan) proposed a concept on dynamic single-mode (DSM) lasers which has led to the development of DFB and DBR lasers for high-speed lightwave systems.[83]

With these new developments, another important industrial field, laser disks, appeared. Optical audio disks or so-called compact discs (CDs) were first marketed around 1984. The monthly production of laser chips has now surpassed 1 million. One might say that the fourth stage started with the requirement for higher-performance lasers in order to develop more sophisticated optical communications, larger-capacity optical disks, and many other applications. For these new lasers, visible lasers with various functions and coherence control are important factors.[84–105]

5.1.2. Fundamentals of Semiconductor Lasers

Semiconductor lasers can operate with applied voltages on the order of 2 V and achieve intensity modulation up to several gigahertz by controlling the injection current. However, since the band-gap energy changes with temperature, the lasing wavelength varies at a rate of 2–3 Å/K. Therefore, some automatic control is required to achieve a constant wavelength. The size of the laser chips is approximately 200 × 250 × 200 μm, corresponding to a cavity length which is 1/100th to 1/10,000th that of gas lasers. Semiconductor laser devices have not become essential for optical communications and optoelectronic systems.

In a ternary-compound semiconductor such as $Ga_{1-x}Al_xAs$, the band-gap energy and refractive index change with the change in the mole fraction, x. The double heterostructure is formed by crystal growth of layers with different values of x. From these types of devices, a continuous output of more than several milliwatts can normally be obtained from red to the near infrared as far as approximately 0.9 μm. Some devices have lifetimes exceeding 10,000 hr. The most popular optical disc (CD) laser is the GaAlAs, which has a wavelength of 0.78 μm, an output of 5 mW, and a lifetime of approximately 8000 hr. The transverse mode is stabilized so that the output is a single transverse mode with divergence angles of 40° × 10° from a near-field pattern of 1 × 5 μm. The CD laser is designed to oscillate with multilongitudinal modes, having a wavelength separation of $\Delta\lambda = 3$ Å. Single wavelength oscillation is necessary for some applications which use the coherent property of the light. If a circular output beam is desired, a narrow stripe laser can be designed for this purpose.

The semiconductor laser is also utilized as a light source for optical fiber communication. The transmission loss of the optical fiber is relatively low in the wavelength band of 1.3 to 1.6 μm (there is a transmission loss minimum of 0.154 dB/km at 1.55 μm). Around 1976, basic research on quaternary compound semiconductor lasers grown on an InP substrate was started in order to match the 1.55-μm low-loss band of silica fibers. With current successes in the growth technique of quaternary crystals for this wavelength range, the integration of photonic devices in this spectral region is now being realized.

5.2. MATERIALS FOR SEMICONDUCTOR LASERS

The emission wavelength of a semiconductor laser nearly corresponds to the band gap of the material, and covers the range from 0.5 to 10 μm. Figure 5.1 shows the wavelength regions of typical laser materials and their applications. Figure 5.2 shows the lattice constant versus band gap for III-V and other important semiconductor materials. For a semiconductor laser, a heterostructure is usually used. The lattice constant shown in Fig. 5.2 is very important to consider for crystal growth because lattice constant matching is normally required to grow one layer on top of another.

5.2.1. Crystals for 1-μm-Band Semiconductor Lasers

5.2.1.1. Importance of the 1-μm Band

In 1967 it was shown by Kao *et al.* that the absorption loss of silicon fiber based on optical loss measurement could be significantly reduced.[106] This was demonstrated by Maurer in 1974 with the development of the silicon fiber formation technique using a chemical vapor deposition (CVD) method.[107] In the early stages of the development of optical fiber communications, the GaAlAs laser (0.85 μm band) was selected. GaAlAs devices with

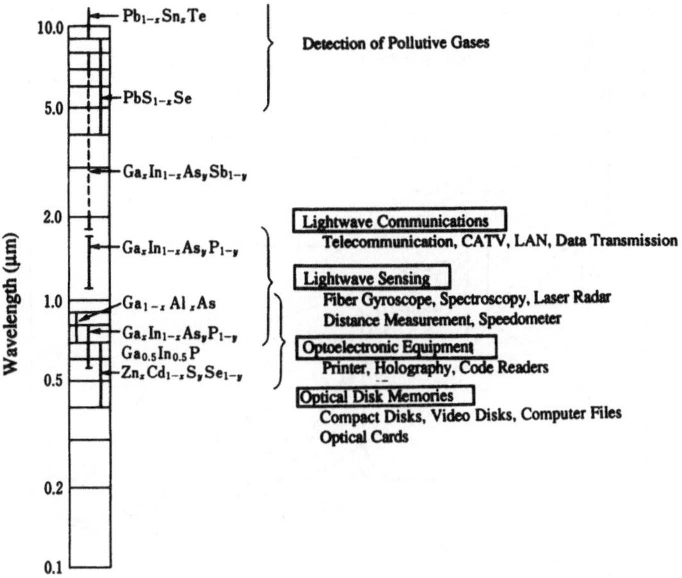

Figure 5.1. Wavelength region of typical laser materials and their applications.

high reliability were fabricated and the so-called first era of optical communication systems was established.

Since 1977, GaInAsP lasers with a wavelength of 1.3 μm have been developed to take advantage of low fiber transmission losses (0.3 dB/km) and zero dispersion at this wavelength. It was the demonstration of the ultralow-loss silicon fiber with a transmission loss of 0.5 dB/km at the 1.3 μm wavelength by Osanai and Horiguchi in 1976 that gave impetus to the research on optical communication in this wavelength range.[108] Also, almost at the same time, Gambling and Payne showed the existence of zero dispersion around 1.3 μm.[109] That characteristic provides for the very low distortion of optical signals when transmitted over long distances.

The characteristics of the 1.3-μm-wavelength optical fiber communication system which one might characterize as the second-generation system, are summarized as follows:

1. Capable of very low transmission loss: Figure 5.3 shows the transmission loss and dispersion of silica fibers. Note that the dispersion goes to 0 at approximately 1.3 μm. The loss continues to decrease in the infrared and a loss of 0.2 dB/km was achieved at the wavelength region of 1.5 μm by the end of 1977 (NTT), and after that, 0.15 dB/km was obtained which is close to the low-loss limit. At shorter wavelengths, Rayleigh scattering is the main cause of light loss, although

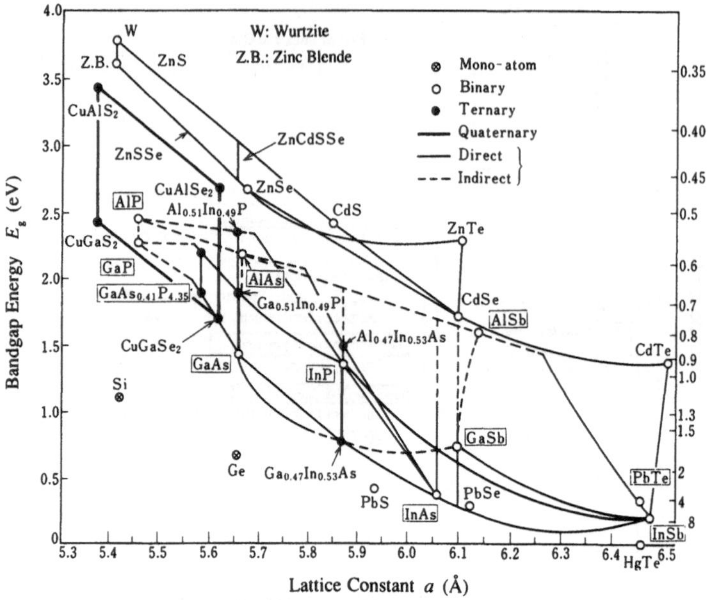

Figure 5.2. Lattice constant versus band gap of III-V semiconductors.

it scales in proportion to λ^{-4} so that it becomes smaller at the long-wavelength region.

2. Wide low-loss band: Since the transmission loss becomes 1 dB/km or less in the wavelength region between 1.1 and 1.6 μm, a wavelength multiplexing system which transmits a number of wavelengths on a single fiber can be utilized.
3. Small material dispersion: Although zero dispersion occurs at 1.3 μm, even in the 1.3 to 1.6 μm region, the dispersion is relatively small, on the order of ± 3 psec/km (distance)·Å (linewidth).
4. A relatively large core diameter: In Fig. 5.3, we show the core diameter which satisfies the single-mode transmission condition for an optical fiber. Single-mode transmission is desirable for coupling and interconnection.

The transmission loss of silica fibers exhibits a minimum in the wavelength region of 1.55 μm, which is a great advantage for long-distance optical communications such as is required for undersea cables. However, since the material dispersion of normal fibers is rather large, the spectral width of the light source must be small when it is modulated. For this reason, a dynamic single-mode laser must be used. Optical communication devices utilizing the 1.55 μm regime can be called third-generation systems.

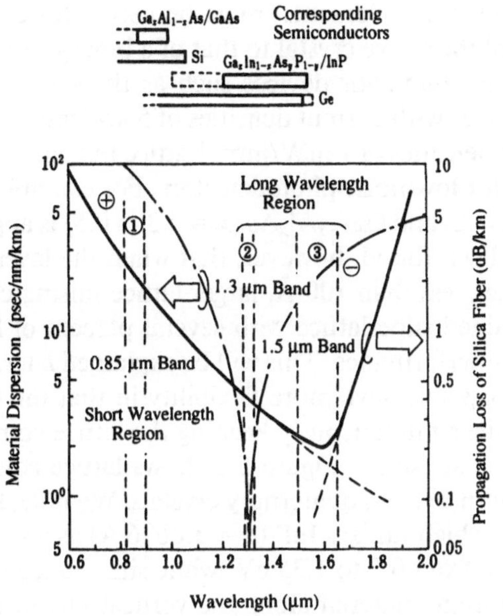

Figure 5.3. Transmission loss and dispersion of optical fibers (silica).

At longer wavelengths, particularly beyond 2 μm (as is shown in Fig. 5.3), the transmission loss of silica fibers increases because of intrinsic infrared absorption related to the Si–O molecular vibration. There have been extensive studies on new optical fiber materials characterized by intrinsic vibration absorption located at wavelengths > 10 μm. In the near future it is expected that optical communications in these longer-wavelength regions between 2 and 10 μm will become possible. As the wavelength becomes longer, the photon energy $h\nu$ (h is Planck's constant and ν is the frequency) becomes small; for example, the energy is 0.12 eV at a wavelength of 10 μm. While the thermal energy kT (k is the Boltzmann constant and T is the absolute temperature) is only 0.03 eV at 70°C, the peak of the blackbody radiation at that temperature is approximately 9 μm, so thermal radiation cannot be ignored in these wavelength regions. As a consequence, coherent optical communications are essential in the wavelength region > 2 μm, and these technologies may be called the fourth and fifth generations of optical communications.

5.2.1.2. Semiconductor Materials

Since binary semiconductor crystals with arbitrary band gaps in the 1 μm region are not available, ternary crystals (consisting of three elements) or quaternary crystals (four elements) must be used. However, matching the lattice constant of the active crystal to that of a binary substrate is normally important for semiconductor devices such as the semiconductor laser or light-emitting diodes with current densities of 5 kA/cm^2, 1 μm or greater, or high light output densities of 1 mW/μm^2. Lattice matching is also important for photodiodes for low-noise photodetectors. For example, the lattice mismatch between GaAs and $Ga_{1-x}Al_xAs$ ($x = 0.3$ to 0.5) is approximately 0.03 to 0.05%. It has been found, however, that when the layer thickness is extremely small, e.g., less than 100 Å, larger lattice mismatches can be tolerated. Such a strained superlattice with several percent of lattice mismatch can exhibit better performance. This will be explained later.

Quaternary crystals have more flexibility in that the band gap can be widely varied while simultaneously keeping the lattice completely matched to a binary crystal substrate. Figure 5.2 shows lattice constants and band gaps for binary, ternary, and quaternary crystals. We note, for example, that $Ga_xIn_{1-x}As_yP_{1-y}$, which utilizes InP (a = 5.8696 Å) as a substrate, can have band gaps ranging from 0.7 to 1.35 eV, while simultaneously matching the lattice of the substrate material along the vertical line in the figure. Other compound crystals correspond at the region of 1 μm band gaps as follows:

$Ga_xIn_{1-x}As_yP_{1-y}$ (InP): $0.92 < \lambda_g < 1.67$ μm
$Ga_{1-x}Al_xAs_ySb_{1-y}$ (GaSb): $0.8 < \lambda_g < 1.7$ μm
$Ga_xIn_{1-x}As_ySb_{1-y}$ (InAs): $1.68 < \lambda_g < 2$ μm
$Ga_xIn_{1-x}As_ySb_{1-y}$ (GaSb): $1.8 < \lambda_g < 2$ μm

where the compound in parentheses is the substrate to be used. Laser-quality crystals are obtained only with lattice mismatches < 0.01% relative to the substrate. The heterojunction composed of $Ga_xIn_{1-x}As_yP_{1-y}$ and InP is mainly used for lasers for photodiodes since it can be constructed so that the carriers are confined as well as the optical mode.

5.2.2. Crystals for Visible to Near-Infrared Semiconductor Lasers

5.2.2.1. The Importance of Visible to Near-Infrared Region

The importance of the visible to the near-infrared region, i.e., the short-wavelength region, can be summarized as follows:

5.2.2.1a. Diffraction Limit of Light. The diameter of the focal spot at the focus of a lens is given by the following equation (note that this equation has appeared two or three times earlier in the text):

$$\Delta D = 1.22\lambda/\mathrm{NA} \tag{5.1}$$

where λ is the wavelength, and NA is the numerical aperture of the focusing lens. For example, the spot size, ΔD, is equal to 1.9 μm when the wavelength is 0.78 μm and the numerical aperture is 0.5. Shorter wavelengths allow tighter focusing with the same numerical aperture.

5.2.2.1b. Sensitivity of Photosensitive Materials. The sensitivity of most photosensitive materials is greatly increased at wavelengths < 0.7 μm; thus, a laser with a short wavelength is desired for such applications as printers and image processing.

5.2.2.1c. Visual Effect. The sensitivity of the human eye ranges between the wavelengths of 0.4 and 0.8 μm, that is, approximately one octave. The highest sensitivity occurs at 0.555 μm or green. It is important to develop lasers in this spectral regime for visual applications.

5.2.2.1d. Color Display. Lasers with three primary colors, i.e., red (R), green (G), and blue (B), are usually utilized for display purposes. Again, short-wavelength lasers are required.

5.2.2.2. Semiconductor Materials

Because many optoelectronic systems require visible lasers, it is important to examine the potential of utilizing semiconductor lasers for these applications. It is generally true that semiconducting materials with large band gaps show a tendency to have small lattice constants, as seen in Fig. 5.2. If the concentration of the Al (x) in the active layer of the $Ga_{1-x}Al_xAs$ system is increased, oscillation at shorter wavelength is obtained. However, if x exceeds 0.4, the minimum value of the conduction band moves from the Γ valley to the X valley. For this reason, the threshold current density starts increasing when λ is less than 0.74 μm. Therefore, in this system, the wavelength of 0.7 μm is a practical limit. A device with $x = 0.5$ emitting at 0.78 μm is commonly used for CDs.

In the region of rather short wavelengths, a double heterostructure of the quaternary $Ga_xIn_{1-x}As_yP_{1-y}$ with GaAs as a substrate, can be utilized. The formation of a cladding using $Al_{1-x}Ga_{1-x}As$ has also been studied. An alternative is an injection-type laser with a large band-gap energy. It is difficult to grow quaternary InGaAlP crystals on a GaAs substrate by liquid-phase epitaxy (LPE). However, metal–organic chemical vapor deposition (MOCVD) can be used to produce a laser with this material which operates at 0.62–0.67 μm.

For any of the III-V crystal types, however, 0.5 μm is the short-wavelength limit. A gallium phosphide (GaP) green light-emitting diode utilizes a transition via impurity levels, since otherwise, the recombination rate of electrons and holes in the indirect transition is very small. Laser oscillation is not possible, however, since it is very difficult to have carriers for recombination only by impurity level transition. In other words, if we wish green or blue semiconductor lasers, II-VI semiconductors such as ZnS, ZnSe, etc., which have band gaps larger than the III-V semiconductors, should be utilized. However, in these II-VI semiconductors, it is hard to dope *p*-type impurities at concentrations larger than 2×10^{18} cm^{-3} because of a self-compensation effect. Densities on this order are required for laser operation. Without sufficient density, the hole injection rate is not large enough and it is difficult to realize a junction laser. One of the advantages of semiconductor lasers is that laser oscillation can be obtained by applying voltages on the order of 1.5 V, which is close to the band gap of the particular semiconductor. Therefore, it is crucial to have a good *p-n* junction. The doping of *p*-type impurities by introducing I-group elements or nitrogens, is a possible approach for making *p-n* junctions in the II-VI semiconductors.

Doping of *p*-type impurities by introducing I-group elements or nitrogens is hopeful for making the *p–n* junction in II–VI semiconductors. In 1991 an injection laser using ZnSe as an active layer cladded by ZnCdSSe was first developed by 3M company.[110] Continuous operation at 77K and pulsed oscillation at room temperature has been demonstrated at blue-green wavelength. As one of the other possible materials for short wavelength band, ZnSe/ZnMgSSe was introduced by Sony, and optically pumped blue lasers operated up to 500 K.[111] Also, injection lasers were fabricated, and pulsed operation at room temperature in blue-green spectral region (λ = 498.5 nm) was demonstrated.[112]

Until we can achieve laser diode operation in the green through blue wavelengths, we will have to rely on second-harmonic generation in order to develop compact light sources in this region.

5.3. BASIC CONCEPT OF SEMICONDUCTOR LASERS

5.3.1. Oscillation Conditions

Under steady-state laser oscillation conditions, light makes a round-trip in the laser resonator and a standing wave is formed, corresponding to a

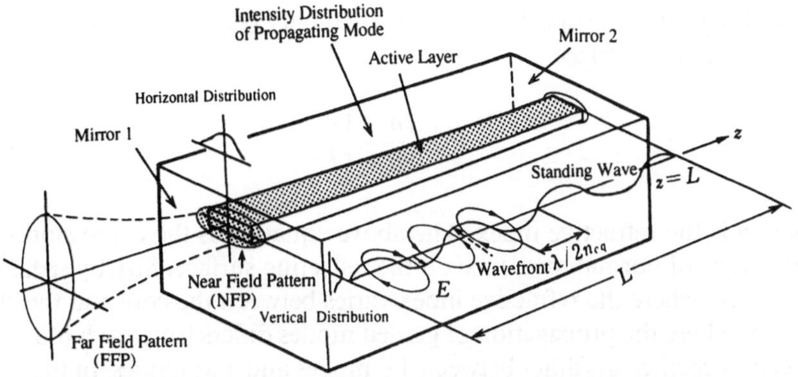

Figure 5.4. Standing wave inside and radiated light of a semiconductor laser.

transverse mode with a wave front which is parallel to the reflecting mirror surfaces. Some portion of the stored light energy in the oscillator cavity is coupled out by transmission through one of the reflecting mirrors.

Figure 5.4 shows a standing wave inside such a resonator cavity. Let us assume that the absorption loss for the optical wave in the semiconductor per unit length is α_a (cm^{-1}), and that the electric field reflectivities and phase shifts of the mirrors are R_1 and R_2, and ϕ_1 and ϕ_2, respectively. When the mode is amplified by induced or stimulated emission, the rate of amplification is expressed by the power gain, g (cm^{-1}). Light starting from $z = 0$ is reflected at $z = L$ and returns again to $z = 0$. From the self-consistent field requirement, the resonance condition may be written as:

$$\exp[-\alpha_a L + gL - j2\beta L - j\phi_1 - j\phi_2]\sqrt{R_1}\sqrt{R_2} = 1 \tag{5.2}$$

where β denotes a propagation constant. Here, the gain that satisfies this resonance condition is called the threshold gain. The threshold gain, g_{th}, is derived from the real part of Eq. (5.2):

$$g_{th} = \alpha_a + \frac{1}{2L}\ln\left(\frac{1}{R_1 R_2}\right) \tag{5.3}$$

The first term, α_a, on the right-hand side of the above equation represents the absorption of the laser light in the medium and is primarily the result of absorption by free carriers. α_a has a typical value of about 10 cm^{-1} in GaAs. The second term on the right-hand side arises from the mirror loss. When the mirror is composed of a plane surface of a dielectric medium,

Fresnel reflection occurs. The Fresnel reflectivity between a dielectric material and air is written as:

$$R = \left(\frac{n-1}{n+1}\right)^2 \tag{5.4}$$

where n is the refractive index. The above equation is the expression for a plane wave of normal incidence on the reflecting surface. Strictly speaking, in the case where the refractive index varies between the core and the cladding, or where the propagation of guided modes differs from each other, the reflection coefficients differ between TE modes and TM modes. In that case, one must use the equivalent refractive index, n_{eq}, associated with a guided mode instead of n.

From the imaginary part of Eq. (5.2), we have the resonance condition:

$$2\beta L + \phi_1 + \phi_2 = 2\pi q \tag{5.5}$$

where q is an integer. Equation (5.5) also applies to the case where reflecting mirrors are coated at the ends of the cavities, or where diffraction gratings are added. If $\phi_1(\lambda)$, $\phi_2(\lambda)$, and $\beta = 2\pi n_{eq}/\lambda$ are substituted into Eq. (5.5), we arrive at the expression:

$$4\pi n_{eq} L/\lambda + \phi_1(\lambda) + \phi_2(\lambda) = 2\pi q \tag{5.6}$$

As the bias current is increased, the gain increases until one of the modes satisfies the threshold oscillation condition and lasing begins. Above that particular bias current, the output power increases rapidly. Table 5.2 compare gain-guided and index-guided lasers. The point where the output power increases rapidly is called the threshold point, and the corresponding current is called the threshold current. The threshold current divided by the area of the active region is called the threshold current density, which is about 1–2 kA/cm^2 in bulk GaAlAs/GaAs or GaInAsP/InP lasers.

5.3.2. Gain Width and Oscillation Spectra

In semiconductor lasers, the gain spectrum has a rather wide spectral width on the order of 300 Å. This linewidth comes from the carrier distribution in the conduction band filling the energy states over a range of energies up to kT, so that when the recombination occurs, leading to gain, it is between holes and electrons having this spread of energies corresponding to the

Table 5.2. Comparison of Gain-Guided and Index-Guided Lasers

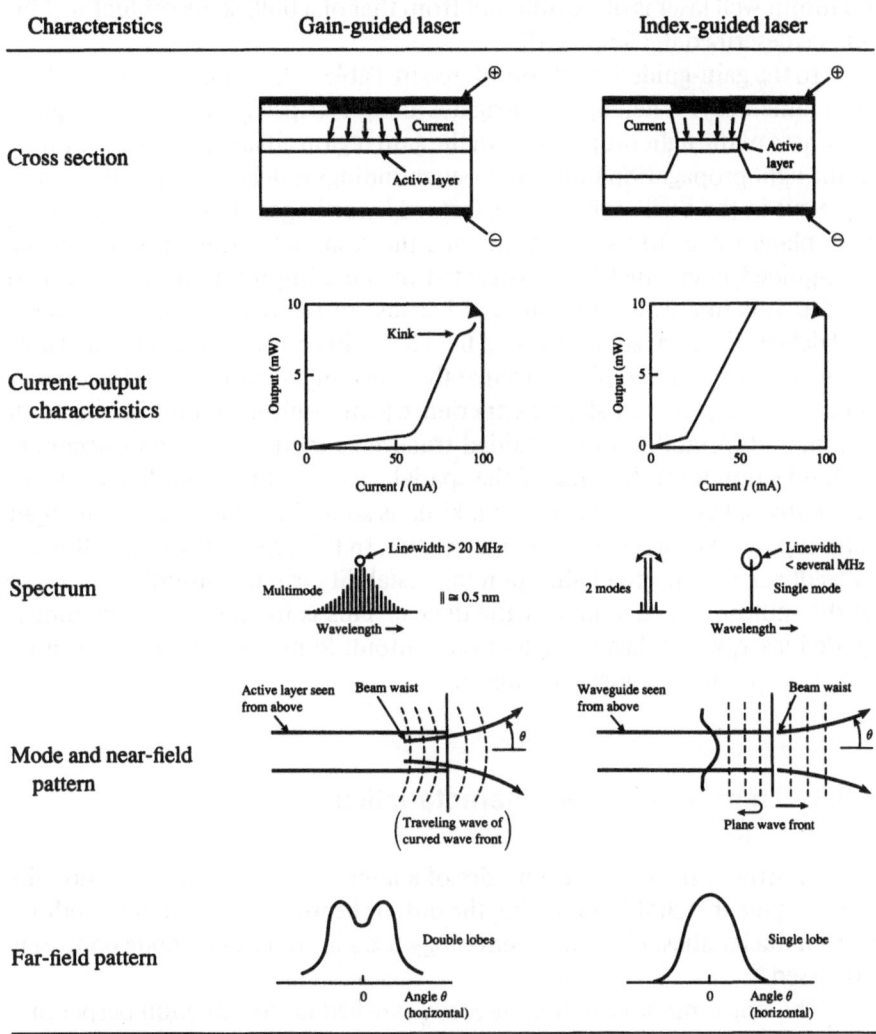

<table>
<tr><th>Characteristics</th><th>Gain-guided laser</th><th>Index-guided laser</th></tr>
<tr><td>Cross section</td><td>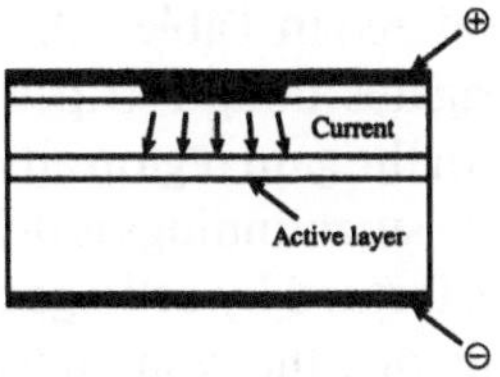
</td><td>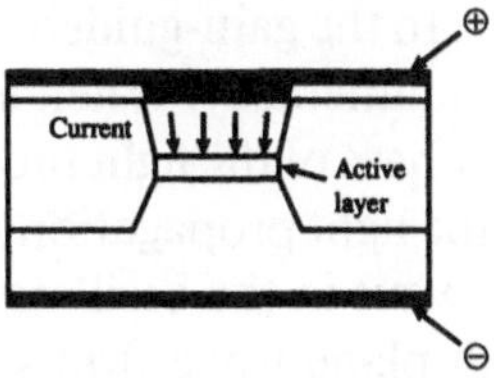
</td></tr>
<tr><td>Current–output characteristics</td><td>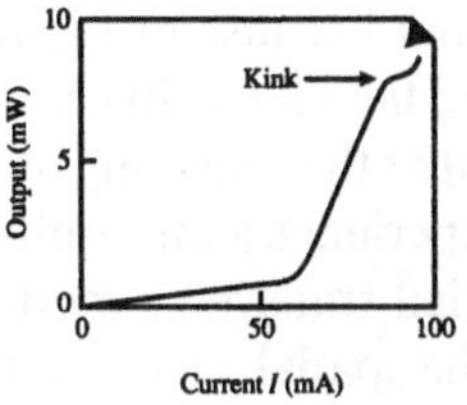
</td><td>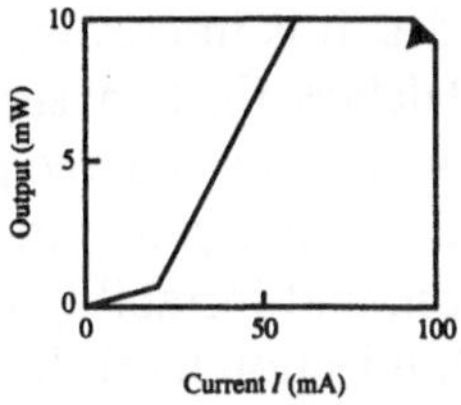
</td></tr>
<tr><td>Spectrum</td><td>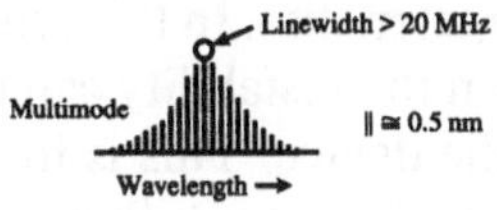
</td><td>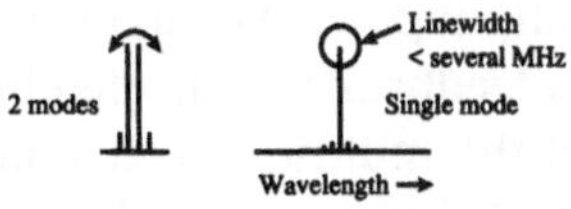
</td></tr>
<tr><td>Mode and near-field pattern</td><td>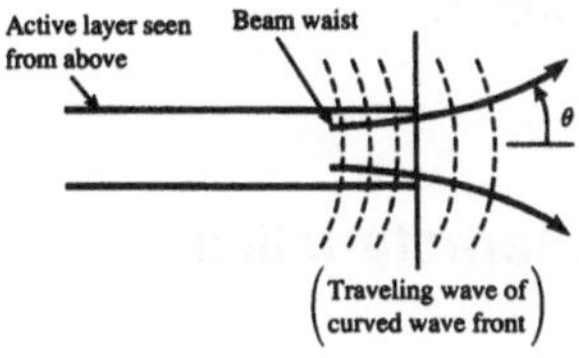
</td><td>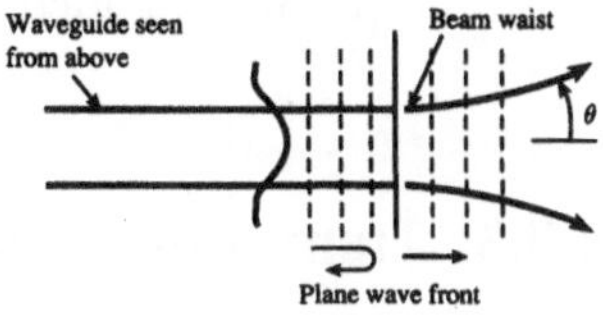
</td></tr>
<tr><td>Far-field pattern</td><td>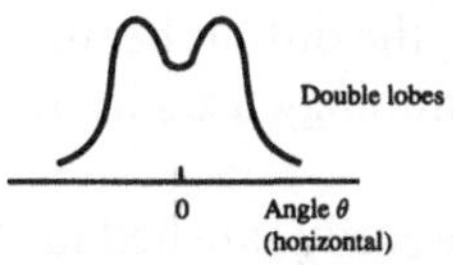
</td><td>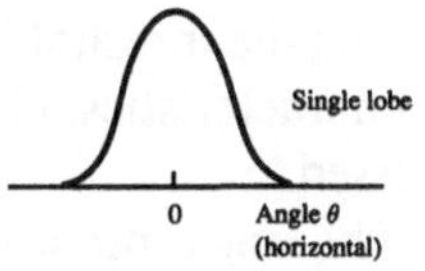
</td></tr>
</table>

gain width. Basically, the gain width is determined by the state density and distribution function of the carriers. The state density expresses the number of states available to the carriers and is given by a parabolic function of the wave number. The carrier distribution function is determined both by the temperature and by the location of the quasi-Fermi level relative to the conduction band. The optical gain of a semiconductor laser is mainly determined by the energy distribution of the electrons injected into the conduction band, since the effective mass of electrons in the conduction band is smaller than that of holes in the valence band. The gain spectrum of a quantum well laser is quite different from that of a bulk semiconductor. This will be described in Chapter 10.

In the gain-guided laser considered in Table 5.2, various modes oscillate simultaneously. When light propagates along a striped geometry gain region, the phase of the light propagating in the gain region advances, while the phase of the light propagation in the lossy surroundings is delayed. Since the standing wave in the oscillator cavity is formed by orthogonal propagating modes with plane wave fronts at the mirrors, the resultant resonant waves in the gain-guided laser must be constructed by superimposing multiple normal modes. It is interesting to note that because of the relative phase advances and delays, the total superimposed field looks like the one indicated in Table 5.2 which has a curved phase front at the reflecting mirrors. Therefore, many modes in the gain-guided laser experience gain resulting in multimode oscillation, and instability in the guided transverse modes sometimes occurs at high injection levels because of the spatial variation of carrier density. Consequently, a bend, which is called a kink, is sometimes observed in the light output power versus current characteristics. In this type of laser the influence of feedback by reflected light upon the instability can be minimized because of the multimode character of the device. This is in contrast to the index-guided laser, which has a single-mode output. Some explanations given by Petermann[37] are included in Table 5.2.

5.3.3. Transverse Mode Characteristics

Control of the transverse modes of a laser is a very effective method for reducing the threshold, stabilizing the output beam, and improving modulation characteristics. Here, the technology used in transverse-mode control is discussed.

The single-mode condition is easily satisfied in the direction perpendicular to the junction since the active layer is very thin, i.e., $<0.1\ \mu m$. This is particularly the case for double heterostructure devices where the normalized index difference, $\Delta = 0.008$, and the active layer thickness for single-mode guiding conditions is thinner than several tens of microns. In the direction parallel to the junction, the mode is controlled by forming a dielectric waveguide with an index difference in the horizontal direction, and making the waveguide effect larger than the gain change caused by the injected current.

Table 5.2 also shows an index-guided laser where the transverse mode has been controlled in this manner.[7] Since the transverse mode character is primarily determined by the index difference and the guide width, the mode

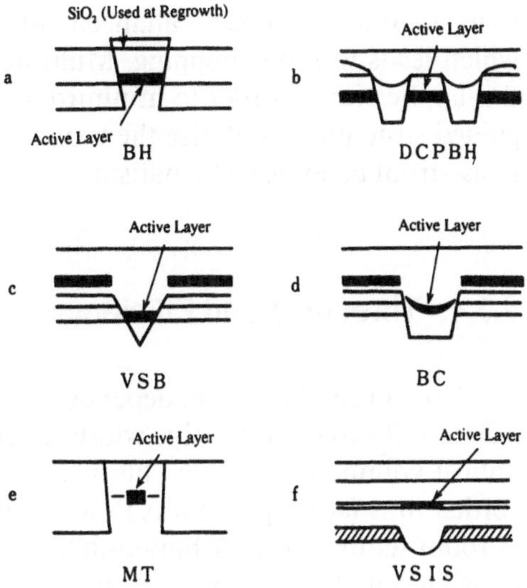

Figure 5.5. Stripe structures for semiconductor lasers. BC, buried crescent type; BH, buried heterostructure; DCPBH, double channel planar buried heterostructure; MT, mass transport buried type; VSB, V-grooved substrate buried type; VSIS, V-grooved substrate internal stripe type.

instability resulting from a gain change can be made very small, and the power output versus injected current increases linearly. As a result, single transverse mode oscillation is possible and the spectral output is very narrow, but the performance of this device is very sensitive to feedback from reflected light. Figure 5.5 shows the three types of index waveguides:

- Buried type: A double heterostructure wafer which has been grown in advance, is processed to form a thin raised mesa line by etching. The side walls are then buried with semiconducting material having a small refractive index through a second crystal growth.
- Selective growth type: A double heterostructure is grown onto a substrate which has been processed into a groove, a terrace, or a mesa in advance, and an active layer and waveguide structure are formed later.
- Distributed impurity type: The index difference for waveguiding is provided through an impurity density variation in the horizontal direction.

Most of the semiconductor lasers which are used in practical systems requiring higher coherence are index-guided lasers. This is essential because single-mode operation is a critical feature. However, the gain for the fundamental mode decreases with an increasing output power. Thus, at high outputs, higher-order modes begin to oscillate. Consequently, it is an important challenge for laser designers to extend the single-mode operation to higher output levels. On the other hand, optical disks and other applications require resistivity to intensity noise enhancement caused by reflected light feedback which leads to mode hopping. Multimode, gain-guided lasers are used for this application in order to minimize the feedback noise by using the gain-guided structure to stabilize the transverse mode and to reduce asymmetric phase-front bending (astigmatism).

5.3.4. Threshold and Efficiency

The threshold current depends on the length of the laser resonator, the mirror reflectivity, the stripe width, the resonator loss, and the ratio of the optical volume to the gain volume, etc. The known threshold currents recorded thus far range from 0.35 to 1.5 mA for continuous wave operation at room temperature. A buried-index waveguide, shown in Table 5.2, and excellent current confinement are necessary for obtaining low thresholds. GaAs and GaInAsP lasers with almost optimum designs can reach thresholds as low as 10 to 30 mA.

Ultralow thresholds are being obtained in quantum well laser devices which have an ultrathin active layer. Breaking the milestone of 1 mA by optimizing the structure is of international interest. A device with 0.35-mA threshold was reported in 1990. The possibility of developing quantum well structures with such low thresholds is an important issue for the development of ultralow power consumption devices.

One of the important characteristics of semiconductor lasers is the light output power (L) as a function of the injection current (I), which is called the I–L characteristic. Continuously operating laser diodes (LDs) can provide several milliwatts of output power, or more, from one mirror facet, although there are, of course, some differences among the various types of laser diodes. For optical communications, the required output is approximately 3 to 10 mW. Recently, an output of 200 to 500 mW has been achieved in a single striped laser, 3.2 W from a wide stripe laser and 12.5 W (10) and 120 W from striped laser arrays with up to 40 elements. The output power is limited by the damage to the electrodes, the junction interfaces, the mirror facets, and catastrophic optical damage to the material. The deterioration of the mirror facets is a serious problem for high output power and the facets need to be protected by coating. This is particularly true for the GaAlAs lasers. By operating semiconductor lasers in a pulsed mode, a peak output of several watts can generally be obtained. These performances will be detailed in Chapter 14.

The differential quantum efficiency, η_D, is defined as the ratio of the increase in the number of output photons for a given increase in the number of injected electrons. In other words, it is related to the slope of the I–L characteristic. This quantum efficiency is written as:

$$\eta_D = \frac{dP/\hbar\omega}{dI/e} = \frac{dP}{dI} \cdot \frac{1}{E_g} \tag{5.7}$$

where e, E_g, and $\hbar\omega$ denote the electron charge, the band-gap energy expressed in electron volts, and photon energy, respectively.

If the total output power from both facets is P, we can obtain $\eta_D = 56\%$ for a GaAs laser by substituting the measured slope of the I–L characteristic, $dP/dI = 0.8$ mW/mA, and $E_g = 1.45$ eV into the above equation. In the case of the cw laser diode, slope efficiencies generally range from 40 to 60%, with a maximum measured slope efficiency of 80%.

5.3.5. Near- and Far-Field Patterns

Figure 5.4 shows the light intensity distribution at the edge of the laser resonator. This is called the near-field pattern (NFP). The NFP can be observed by a microscope and an infrared monitor. It is desirable for virtually all applications that the NFP not change significantly with current.

The far-field pattern (FFP) describes the radiated beam shape far from the laser. Usually, it is expressed with plots in terms of divergence angles relative to the central axis. The FFP in the vertical direction is usually larger than that in the horizontal direction. In an ordinary striped laser, a half-angle of 15° is common for the horizontal direction, and 30° for the vertical direction.

The FFP does not always correspond to the Fourier transform of the NFP. This is only the case if the laser mode exiting the semiconductor cavity is a plane wave, so the phase variation across the NFP can be ignored. In the case of the gain-guided laser, however, the wave front at the output mirror is not always a plane wave. Particularly in the horizontal direction, the wave front is convex seen from outside the laser, and appears to emerge from a point source within the laser device at a finite distance from the exit facet. In the vertical direction, however, the wave front is plane, because of the thin double heterostructure which leads to good waveguiding properties. Consequently, the output of gain-guided laser devices has an astigmatism.[7]

5.3.6. Temperature Characteristics

Semiconductor laser performance is relatively sensitive to temperature. The threshold current, I_{th}, and the differential quantum efficiency, η_D, degrade with temperature. The temperature dependence of the threshold current is written as:

$$I_{th}(T) = I_{th}(T') \exp[(T - T')/T_o] \tag{5.8}$$

where $I_{th}(T)$ is the threshold at T and T_o is called a characteristic temperature. T' is the reference temperature, usually approximately 23°C. $T_o = 120$ to 150 K in GaAs/GaAlAs lasers, and $T_o = 50$ to 70 K in GaInAsP/InP lasers. A larger characteristic temperature is preferable to minimize the effects of temperature variation. The causes of a low characteristic temperature in the GaInAsP/InP system are discussed in an article by Suematsu[40] and are summarized as follows:

1. Auger recombination
2. Intra-valence-band absorption
3. Hot electron effects

PROBLEMS

5.1. State the importance of lattice matching for epitaxial crystal growth. What, on the other hand, is the significance of highly mismatched hetero-epitaxy?

5.2. Describe possible compound semiconductors lattice-matched to GaAs and InP, respectively.

5.3. What is a suitable material for lasers and detectors in the 1.55-μm wavelength band where the silica fiber exhibits minimum transmission loss?

5.4. What are the important applications of a short-wavelength semiconductor laser?

5.5. Explain the role of the cladding layers which sandwich the active layer in semiconductor lasers. (Consider optical confinement and carrier confinement)

5.6. What is the shortest wavelength obtainable by III-V compound semiconductors?

5.7. Explain why the far-field pattern of a semiconductor laser is like that shown in Fig. 5.5 originating from a horizontal ellipsoid.

5.8. Show that the threshold gain is expressed by Eq. (5.3). Calculate the threshold gain when $\alpha_a = 10\ \text{cm}^{-1}$, $R_1 = 30\%$, $R_2 = 95\%$, and $L = 100\ \mu\text{m}$. Hint: The gain factor for the electric field is 1/2 of the power gain.

5.9. Why do the Fresnel reflectivities differ for the TE and TM modes?

5.10. What is the phase shift suffered when light incident from the air is reflected at the surface of a dielectric medium?

REFERENCES

Initial Demonstration

1. M. I. Nathan, W. P. Dumke, G. Burns, F. H. Dill, Jr., and G. Lasher, *Appl. Phys. Lett.* **1,** 62 (Feb. 1, 1962).
2. R. N. Hall, G. H. Fenner, J. D. Kingsley, T. J. Soltys, and R. D. Carlson, *Phys. Rev. Lett.* **9,** 366 (1962).
3. T. M. Quist, R. H. Rediker, R. J. Keyes, W. E. Krag, B. Lax, A. L. McWhorter, and H. J. Zeiger, *Appl. Phys. Lett.* **1,** 91 (Feb. 15, 1962).
4. N. Holonyak, Jr., and S. F. Bevacqua, *Appl. Phys. Lett.* **1,** 82 (Feb. 1962).
5. R. Braunstein, *Phys. Rev.* **99**(9), 1982 (Sept. 1955).
6. J. Nishizawa and Y. Watanabe, Japanese Patent 273217 (1957).
7. N. G. Basov, B. M. Vul, and Y. M. Popov, *Zh. Eksp. Teor. Fiz.* **3,** 587 (Aug. 1959).
8. H. C. Casey and M. B. Panish, *Heterostructure Lasers,* Academic Press, New York (1978).

Early Research Stage

9. G. Lasher and F. Stern, *Phys. Rev.* **133A,** 553 (Jan. 1964).
10. M. H. Pilkuhn and H. Rupprecht, *Solid-State Electron.* **7,** 905 (1964).
11. Y. Nannichi, *J. Appl. Phys.* **36**(4), 1499 (April 1965).
12. W. W. Anderson, *IEEE J. Quantum Electron.* **QE-1,** 228 (Sept. 1965).
13. B. S. Goldstein *et al. Proc. IEEE* **53**(2), 195 (Feb. 1965).
14. T. Ikegami and Y. Suematsu, *Proc. IEEE,* **55**(1), 122 (1967).
15. J. Takamiya, F. Kitasawa, and J. Nishizawa, *Proc. IEEE* (Lett.) **56,** 135 (Jan. 1968).
16. T. Ikegami and Y. Suematsu, *Trans. IECE Jpn.* **51-B,** 57 (Feb. 1968); *IECE J. Quantum Electron.* **QE-4,** 148 (April 1968).
17. J. C. Dyment, *Appl. Phys. Lett.* **10**(3), 84 (Feb. 1967).
18. W. Susaki, *Jpn. J. Appl. Phys.* **6**(8), 977 (Aug. 1967).
19. H. Kressel and H. P. Mierop, *J. Appl. Phys.* **38**(12), 5419 (Dec. 1967).
20. H. Haug, *Phys. Rev.* **184**(2), 338 (Feb. 1969).
21. N. G. Basov, V. V. Nikitin, and A. S. Semenov, *Usp. Fiz. Nauk* **97**(4), 561 (April 1969). [*Sov. Phys. Usp.* **12,** 219 (Sept.–Oct. 1969)].
22. H. Kressel and H. Nelson, *RCA Rev.* **30,** 106 (March 1969).
23. J. E. Ripper and T. L. Paoli, *Appl. Phys. Lett.* **18,** 466 (May 1971).
24. T. Ikegami, K. Kobayashi, and Y. Suematsu, *Trans. IECE Jpn.* **53-B**(5), 53 (May 1970); T. Ikegami and Y. Suematsu, *Trans. IECE Jpn.* **53-B**(9), 513 (Sept. 1970).
25. T. P. Lee and R. H. R. Roldan, *IECE J. Quantum Electron.* **QE-6,** 339 (June 1970).
26. C. J. Hwang, *Phys. Rev.* **B2**(1), 4117, 4126 (Nov. 1970).
27. H. Kressel and H. Nelson, *RCA Rev.* **30,** 106 (1969).
28. H. Kroemer, *Proc. IEEE* **51**(12), 1782 (Dec. 1963).
29. W. Susaki, T. Sogo, and T. Oku, *IEEE J. Quantum Electron.* **QE-4,** 422 (June 1968).
30. H. C. Casey and M. B. Panish, *J. Appl. Phys.* **40**(12), 4190 (Nov. 1969).
31. Z. I. Alferov, V. M. Andreev, E. L. Portnoi, and M. K. Trukan, *Fiz. Tekh. Poluprovodn.* **3,** 1328 (Sept. 1969). [*Sov. Phys. Semicond.* **3,** 1107 (March 1970)]
32. I. Hayashi, M. B. Panish, P. W. Foy, and A. Sumski, *Appl. Phys. Lett.* **17**(3), 109 (Aug. 1970).

33. I. Hayashi, M. B. Panish, and F. K. Reinhart, *J. Appl. Phys.* **42**(4), 1929 (April 1971).
34. T. Ikegami, *IEEE J. Quantum Electron.* **QE-8,** 470 (1972).
35. M. B. Panish, H. C. Casey, Jr., S. Sumski, and P. W. Foy, *Appl. Phys. Lett.* **22,** 590 (June 1973).
36. Y. Suematsu and M. Yamada, *Trans. IECE Jpn. (C)* **57-C**(11), 434 (1975).
37. K. Petermann, *7th Eur. Conf. Opt. Commun.,* No. 10.1 (Sept. 1981).
38. W. T. Tsang, R. A. Logan, and J. A. Ditzenberger, *Electron. Lett.* **18,** 845 (1982).
39. D. R. Scifres, C. Lindstrom, R. D. Burnham, W. Streofer, and T. L. Paoli, *Electron. Lett.* **19,** 169 (March 1983).
40. Y. Suematsu, *Proc. IEEE* **71,** 692 (June 1984).
41. I. Hayashi, *IEEE Trans. Electron Devices* **ED-31**(11), 1630 (1984).

Long-Wavelength Lasers

42. A. P. Bogatov, L. M. Dolginov, P. G. Eliseev, M. G. Milvidskii, B. N. Sverdlov, and E. G. Shevchenko, *Sov. Phys. Semicond.* **9**(10), 1282 (Oct. 1975).
43. J. J. Hsieh, J. A. Rossi, and J. P. Donnelly, *Appl. Phys. Lett.* **28,** 709 (June 1976).
44. K. Oe and K. Sugiyama, *Jpn. J. Appl. Phys.* **15**(12), 740 (Dec. 1976).
45. T. P. Pearsall, B. I. Miller, R. J. Capik, and K. J. Bachmann, *Appl. Phys. Lett.* **28,** 499 (May 1976).
46. T. Yamamoto, K. Sakai, S. Akiba, and Y. Suematsu, *Electron. Lett.* **13,** 142 (March 1977).
47. K. Oe, S. Ando, and K. Sugiyama, *Jpn. J. Appl. Phys.* **16**(7), 1273 (July 1977).
48. Y. Itaya, Y. Suematsu, and K. Iga, *Jpn. J. Appl. Phys.* **16**(6), 1057 (June 1977).
49. K. Wakao, K. Moriki, T. Kambayashi, and K. Iga, *Jpn. J. Appl. Phys.* **16**(11), 2073 (Nov. 1977).
50. Z. I. Alferov, A. T. Gorelenok, P. Kopiev, V. N. Mdivani, and V. K. Tibilov, *Pisma Zh. Tekh. Fiz.* **3**(22), 1169 (1977).
51. M. A. Pollack, R. E. Nahory, J. C. DeWinter, and A. A. Ballman, *Appl. Phys. Lett.* **33**(4), 314 (Aug. 1978).
52. T. Yamamoto, K. Sakai, S. Akiba, and Y. Suematsu, *IEEE J. Quantum Electron.* **QE-14,** 95 (Feb. 1978).
53. Y. Itaya, Y. Suematsu, S. Katayama, K. Kishino, and S. Arai, *Jpn. J. Appl. Phys.* **18**(9), 1795 (Sept. 1979).
54. S. D. Hersee, A. C. Carter, R. C. Goodfellow, G. Hawkins, and I. Griffith, *Solid-State Electron Devices* **3**(6), 179 (Nov. 1979).
55. R. J. Nelson, *Appl. Phys. Lett.* **35**(9), 654 (Nov. 1979).
56. R. E. Nahory and M. A. Pollack, *Electron. Lett.* **14,** 727 (1978).
57. T. Yamamoto, K. Sakai, and S. Akiba, *IEEE J. Quantum Electron.* **QE-15,** 684 (1979).
58. K. Mizuishi, M. Hirao, S. Tsuji, H. Sato, and M. Nakamura, *Jpn. J. Appl. Phys.* **19**(7), L429 (July 1980).
59. H. Imai, M. Morimoto, H. Ishikawa, K. Hori, and M. Takusagawa, *Appl. Phys. Lett.* **38,** 16 (Jan. 1981).
60. E. Oomura, T. Murotani, H. Higuchi, H. Namizaki, and W. Susaki, *IEEE J. Quantum Electron.* **QE-17,** 646 (May 1981).
61. A. G. Steventon, R. E. Spillet, R. E. Hobbs, M. G. Burt, P. J. Fiddyment, and J. V. Collins, *IEEE J. Quantum Electron.* **QE-17,** 602 (May 1981).
62. K. Iga and B. I. Miller, *Electron. Lett.* **16**(22), 830 (Oct. 1980).

63. M. Hirao, S. Stujii, K. Mizushima, A. Doi, and M. Nakamura, *J. Opt. Commun.* **1**(1), 10 (Sept. 1980).
64. J. J. Hsieh and C. C. Shen, *Appl. Phys. Lett.* **30**, 429 (April 1977).
65. K. Kishino, Y. Suematsu, and Y. Itaya, *Electron. Lett.* **15**, 134 (Feb. 1979).
66. K. Moriki, K. Wakao, M. Kitamura, K. Iga, and Y. Suematsu, *Jpn. J. Appl. Phys.* **19**(11), 2191 (Nov. 1980).
67. H. Nishi, M. Yano, Y. Nishitani, Y. Akita, and M. Takusagawa, *Appl. Phys. Lett.* **35**, 232 (Aug. 1979).
68. R. J. Nelson, P. D. Wright, P. A. Barnes, R. L. Brown, T. Cella, and R. G. Sobers, *Appl. Phys. Lett.* **36**, 358 (1980).
69. I. Mito, M. Kitamura, and K. Kobayashi, *Technical Digest of Optical Fiber Commun.* Phoenix, ThBB2 (April 1982).
70. K. Iga and B. I. Miller, *Electron. Lett.* **16**(22), 830 (Oct. 1980).
71. T. Horikoshi, T. Kobayashi, and Y. Furukawa, *Jpn. J. Appl. Phys.* **18**(12), 2237 (Dec. 1979).
72. M. Fukuda and K. Wakita, *Jpn. J. Appl. Phys.* **19**(11), L667 (Nov. 1980).
73. D. Botez and G. J. Herskowitz, *Proc. IEEE* **68**(6), 689 (June 1980).
74. M. Nakamura and S. Tsuji, *IEEE J. Quantum Electron.* **QE-17**, 994 (June 1981).
75. P. Marshall, E. Schlosser, and C. Wolk, *Electron. Lett.* **15**(1), 38 (Jan. 1979).
76. S. Akiba, K. Sakai, and T. Yamamoto, *Jpn. J. Appl. Phys.* **17**(10), 1899 (Oct. 1978).
77. S. Arai, Y. Suematsu, and Y. Itaya, *Jpn. J. Appl. Phys.* **18**(3), 709 (March 1979).
78. G. D. Henshall and P. D. Greene, *Electron. Lett.* **15**(20), 621 (Sept. 1979).
79. S. Akiba, K. Sakai, Y. Matsushima, and T. Yamamoto, *Electron. Lett.* **15**, 606 (Sept. 1979).
80. H. Kawaguchi, T. Takahei, Y. Toyoshima, H. Nagai, and G. Iwane, *Electron. Lett.* **15**, 669 (Oct. 1979).
81. S. Arai, M. Asada, Y. Suematsu, and Y. Itaya, *Jpn. J. Appl. Phys.* **18**(12), 2333 (Dec. 1979).
82. I. P. Kaminow, R. E. Nahory, M. A. Pollack, L. W. Stulz, and J. C. DeWinter, *Electron. Lett.* **15**, 763 (Nov. 1979).
83. K. Iga and Y. Suematsu, *1st Eur. Conf. Integrated Opt.*, p. 70 (1981).

Short-Wavelength Lasers

84. R. D. Burnham, N. Holonyak, Jr., D. L. Keune, and D. R. Scifres, *Appl. Phys. Lett.* **18**, 160 (Feb. 1971).
85. Z. I. Alferov, I. N. Arsent'ev, D. Z. Garbuzov, S. G. Konnikov, and V. D. Rumyantsev, *Sov. Tech. Phys. Lett.* **1**, 147 (April 1975).
86. H. Kressel, G. H. Olsen, and C. J. Nuese, *Appl. Phys. Lett.* **30**, 249 (March 1977).
87. S. Mukai, H. Yajima, and J. Shimada, *Jpn. J. Appl. Phys.* **20**, L729 (Oct. 1981).
88. T. Suzuki, I. Hino, A. Gomyo, and K. Nishida, *Jpn. J. Appl. Phys.* **21**, L731 (Dec. 1982).
89. Y. Kawamura, H. Asahi, H. Nagai, and T. Ikegami, *Electron. Lett.* **19**, 931 (March 1983).
90. A. Fujimoto, H. Watanabe, M. Takeuchi, and M. Shimura, *Jpn. J. Appl. Phys.* **23**, L720 (Sept. 1984).
91. A. Usui, T. Matsumoto, M. Inai, I. Mito, K. Kobayashi, and H. Watanabe, *Jpn. J. Appl. Phys.* **24**, L163 (March 1985).
92. K. Kishino, Y. Kaneko, and A. Harada, *Jpn. J. Appl. Phys.* **24**, L358 (May 1985).
93. K. Kobayashi, S. Kawata, A. Gomyo, I. Hino, and T. Suzuki, *Electron. Lett.* **21**, 931 (Sept. 1985).

94. M. Ikeda, Y. Mori, H. Sato, K. Kaneko, and N. Watanabe, *Appl. Phys. Lett.* **47,** 1027 (Nov. 1985).
95. M. Ishikawa, Y. Ohba, H. Sugawara, M. Yamamoto, and T. Nakanishi, *Appl. Phys. Lett.* **48,** 207 (Nov. 1985).
96. J. P. Andre, E. Dupont-Nivet, D. Monori, J. N. Patillon, M. Erman, and T. Ngo, *J. Cryst. Growth* **77,** 354 (1986).
97. K. Nakano, M. Ikeda, A. Toda, and C. Kojima, *Electron. Lett.* **23,** 894 (Aug. 1987).
98. K. Iga, H. Uenohara, and F. Koyama, *Electron. Lett.* **22,** 1008 (1986).
99. K. Kishino, A. Kikuchi, Y. Kaneko, and I. Nomura, *Appl. Phys. Lett.* **58,** 1822 (1991).
100. J. Rennie, M. Okajima, M. Watanabe, and G. Hatakoshi, *13th IEEE Semicond. Laser Conf., Tech. Dig.* No. G-4, p. 158 (1992).
101. T. Kamizono, S. Arimoto, H. Watanabe, K. Kodoiwa, E. Omura, S. Kakimoto, and K. Ikeda, *13th IEEE Semicond. Laser Conf., Tech. Dig.* No. D-30, p. 94 (1992).
102. H. Hamada, K. Tominaga, M. Shono, S. Honda, K. Yodoshi, and T. Yamaguchi, *Electron. Lett.* **28,** 1834 (1992).
103. A. Valster, C. J. van der Poel, M. N. Finke, and M. J. B. Boermanns, *13th IEEE Semicond. Laser Conf., Tech. Dig.* No. G-1, p. 152 (1992).
104. M. A. Haase, J. Qiu, J. M. Depuydt, and H. Cheng, *Appl. Phys. Lett.* **59,** 1272 (1991).
105. M. Ozawa, H. Okujima, Y. Morinaga, F. Hiei, and K. Akimoto, *13th IEEE Semicond. Laser Conf., Tech. Dig.* (Post Deadline Papers), No. PD-16, p. 31 (1992).
106. C. K. Kao and T. W. Davies, *J. Sci. Instrum.,* **1,** 1063 (1968).
107. R. D. Maurer, *Tenth Int'l. Cong. Glass,* **2A,** 10 (1974).
108. M. Horiguchi and H. Osani, *Electron. Lett.,* **12,** 310 (1976).
109. D. N. Payne and W. A. Gambling, *Electron. Lett.,* **11,** 176 (April 1975).
110. M. A. Hasse, J. Qui, J. M. Depuyt, H. Cheng, "Blue-green Laser Diodes," in *Appl. Phys. Lett.,* **59,** 1272–1274 (1991).
111. Y. Morinaga, H. Okuyama, and Akimoto, "Photobumped Blue Lasers in ZnSSe-ZnMgSSe Double Heterostructures Operating up to 500K," in *Conf. on Solid State Device and Materials,* **Post Deadline Papers,** 707–708 (1992).
112. S. Itoh, H. Okuyama, S. Matsumoto, N. Nakayama, T. Ohata, T. Miyajima, A. Ishibashi, and K. Akimoto, "Room Temperature Pulsed Operation of 498nm Laser with ZnMgSSe Cladding Layers," in *Electron. Lett.,* **29,** 9, 766–768 (1993).

SIX

LIGHT BEAMS

We know that the light from a laser propagates as a well-defined beam. This is because the normal mode in the laser resonator which consists of plane mirrors or concave mirrors, is also a well-defined beam, as will be described in Chapter 8. In this chapter, we will discuss optical propagating modes. This discussion includes the development of the normal mode analysis for a distributed index waveguide, which also serves as a good model for semiconductor lasers (see Fig. 6.1). We will also explain diffraction phenomena by a linear combination of these normal modes. Finally, we discuss a waveform matrix and a ray transfer matrix by which the propagation of a Gaussian profile beam in free space is conveniently expressed. The object of this chapter is, therefore, to study laser beam propagation which will be helpful for the discussion about laser resonators in Chapter 8.

6.1. EQUATIONS EXPRESSING AN ELECTROMAGNETIC FIELD OF LIGHT

We begin this section with Maxwell's equations from which a wave equation can be derived which expresses the propagating electromagnetic field of light. We then develop the concept of a normal mode and a beam wave (beam mode) propagating in free space.

Maxwell's equations, written in MKS units, are given as follows:

$$\nabla \times \mathbf{E} = -\mu_0 \frac{\partial \mathbf{H}}{\partial t} \tag{6.1}$$

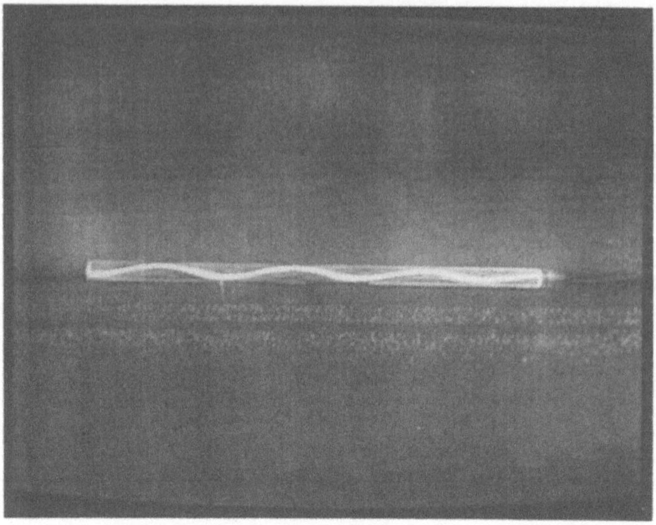

Figure 6.1. Light beams propagating through a distributed-index optical fiber. (An argon laser light beam is propagating through a plastic fiber which was prepared in the author's laboratory.)

$$\nabla \times \mathbf{H} = \mathbf{j} + \frac{\partial \mathbf{D}}{\partial t} \tag{6.2}$$

$$\mathbf{j} = \sigma\mathbf{E} + \mathbf{J} \tag{6.3}$$

$$\mathbf{D} = \varepsilon\mathbf{E} + \mathbf{P} \tag{6.4}$$

$$\nabla \cdot \mathbf{D} = 0 \tag{6.5}$$

$$\nabla \cdot (\mu_0 \mathbf{H}) = 0 \tag{6.6}$$

Here we have assumed that the magnetic permeability of a laser medium is negligible, so the free space value of μ_0 is used. **P** and **J** are the polarization and current densities, respectively, and arise from the dipole transition in the material to be considered. $\sigma\mathbf{E}$ is the conductive current (σ is electrical conductivity) of the medium. $\varepsilon\mathbf{E}$ is the polarization in terms of the background dielectric constant, ε. Consequently, **J** and **P** are two state functions which are induced by the electromagnetic field and contribute to the attenuation or the amplification of light, which is the basic origin for a laser amplification.

From Eq. (6.1) we have:

$$\nabla \times \nabla \times \mathbf{E} = -\mu_0 \frac{\partial}{\partial t}\left[\mathbf{J} + \sigma\mathbf{E} + \frac{\partial}{\partial t}(\varepsilon\mathbf{E} + \mathbf{P})\right] \tag{6.7}$$

From Eqs. (6.4) and (6.5) we have:

$$\nabla \cdot (\varepsilon \mathbf{E} + \mathbf{P}) = \nabla \cdot (\varepsilon E) + \nabla \cdot \mathbf{P} = 0 \tag{6.7a}$$

However, the polarization, $\mathbf{P}$, is an induced phenomenon and exists only when the electromagnetic field associated with propagating light is present. Therefore, its divergence must be 0; $\nabla \cdot \mathbf{P} = 0$. From the above, by using $\nabla \cdot (\varepsilon \mathbf{E}) = \nabla\varepsilon \cdot \mathbf{E} + \varepsilon \nabla \cdot \mathbf{E}$, we have:

$$\nabla \cdot \mathbf{E} = -\frac{\nabla\varepsilon}{\varepsilon} \cdot \mathbf{E} = -\nabla(\ln\varepsilon) \cdot \mathbf{E} \tag{6.7b}$$

By using the well-known vector identity,

$$\nabla \times \nabla \times \mathbf{E} = -\nabla^2 \mathbf{E} + \nabla\nabla \cdot \mathbf{E} \tag{6.7c}$$

and from Eq. (6.7b), Eq. (6.7) is written as:

$$\nabla^2 \mathbf{E} + \nabla\left(\frac{\nabla\varepsilon}{\varepsilon} \cdot \mathbf{E}\right) - \mu_0 \frac{\partial}{\partial t}\left(\sigma + \varepsilon \frac{\partial}{\partial t}\right)\mathbf{E} = \mu_0 \frac{\partial}{\partial t}\left(\mathbf{J} + \frac{\partial}{\partial t}\mathbf{P}\right) \tag{6.8}$$

This equation can be modified according to various cases as shown below.

6.1.1. Passive Case (J = 0, P = 0)

When the quantum system does not exist in the medium, Eq. (6.8) becomes:

$$\nabla^2 \mathbf{E} + \nabla\left(\frac{\nabla\varepsilon}{\varepsilon} \cdot \mathbf{E}\right) - \mu_0 \frac{\partial}{\partial t}\left(\sigma + \varepsilon \frac{\partial}{\partial t}\right)\mathbf{E} = 0 \tag{6.9}$$

In the case where there is no loss in the medium and the gradient of the dielectric constant is not very large (this will be detailed later), the terms involving $\nabla\varepsilon$ and σ can be neglected. If the temporal behavior is assumed to be sinusoidal ($e^{j\omega t}$), the following equation can be obtained:

$$\nabla^2 \mathbf{E} + \omega^2 \mu_0 \varepsilon(r) \mathbf{E} = 0 \tag{6.10}$$

This equation can be used as an approximate wave equation when the dielectric constant has a spatial variation.

On the other hand, if ε is constant and the field has sinusoidal spatial variation ($e^{-jk:r}$), where the magnitude of k is a propagation constant ($k^2 = \omega^2\mu_0\varepsilon$), then $\nabla^2\mathbf{E} = -k^2\mathbf{E}$, and the following equation is obtained:

$$\frac{d^2\mathbf{E}}{dt^2} + \frac{\sigma}{\varepsilon}\frac{d\mathbf{E}}{dt} + \omega^2\mathbf{E} = 0 \tag{6.11}$$

6.1.2. Active Case

Note that $\sigma + \varepsilon(\partial/\partial t)$ on the left side and $J + \varepsilon\partial P/\partial t$ on the right side of Eq. (6.8) both contain derivatives with respect to t. In the case where the temporal behavior is $e^{j\omega t}$, the time derivative $\partial/\partial t$ can be simply replaced by $j\omega$. It is then possible to use equivalent complex parameters, P_{eq} or J_{eq} and ε_{eq} or σ_{eq}:

$$\mathbf{J} + \frac{\partial}{\partial t}\mathbf{P} = \frac{\partial}{\partial t}\mathbf{P}_{eq} \text{ or } \mathbf{J}_{eq} \tag{6.12}$$

$$\sigma + \varepsilon\frac{\partial}{\partial t} = \varepsilon_{eq}\frac{\partial}{\partial t} \text{ or } \sigma_{eq} \tag{6.13}$$

Table 6.1 shows these equivalent parameters denoted by "eq" for various lasers.

Table 6.1. Equivalent Parameters

		$\mathbf{P}_{eq}$		$\mathbf{J}_{eq}$	
	σ	Real part	Imaginary part	Real part	Imaginary part
Gas lasers, solid-state lasers, etc.	Resonator loss	Contribution to dielectric constant	Gain	—	—
Semiconductor lasers	Absorption loss Resonator loss	—	—	Gain	Contribution to dielectric constant

6.2. NORMAL MODES

Here we introduce the concept of normal modes. To be exact, normal modes can exist only when there is no loss in a medium. For example, in a resonator consisting of a dielectric waveguide, the form of a propagating wave is written as $e^{-j\beta z}R(x, y)$. If z is the propagation direction of the light (see Fig. 6.2), then the following wave equation can be derived:

$$\left(\frac{\partial^2}{\partial x^2} + \frac{\partial^2}{\partial y^2}\right)R + [\omega^2\mu_0\varepsilon(x, y) - \beta^2]R = 0 \tag{6.14}$$

According to the distribution, $\varepsilon(x, y)$ of the dielectric constant, the eigenfunction $R(x, y) = u_n(x, y)$, and the eigenvalue β_n^2 can be determined.

On the other hand, in a Fabry–Perot resonator consisting of two reflecting mirrors facing each other, the relationship of normal modes $u_n(x, y)$ and the eigenvalue of the propagation constant, β_n, can be determined from the wave equation (Helmholtz equation) and the boundary conditions. Generally, the eigenfunctions are orthogonal and can be normalized as:

$$\int u_r(x, y) \cdot u_m^*(x, y)dxdy = \delta_{nm} \tag{6.15}$$

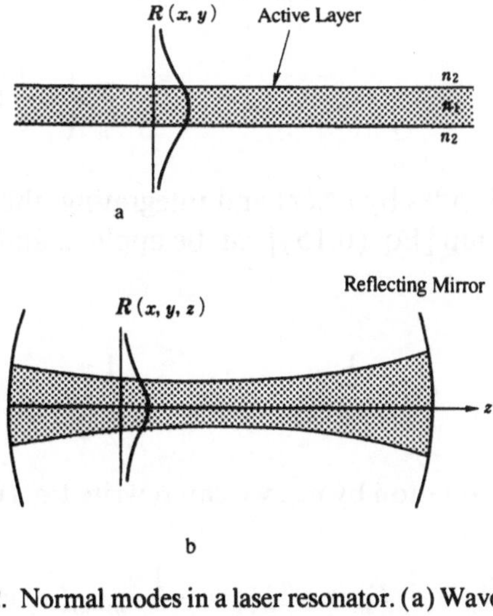

Figure 6.2. Normal modes in a laser resonator. (a) Waveguide-type resonator; (b) open resonator.

Furthermore, the eigenfunctions form a complete set and are written as:

$$\sum_{n=0}^{\infty} u_n(x, y) \cdot u_n^*(x'y') = \delta(x - x')\delta(y - y') \tag{6.16}$$

By using these normal modes, we expand the electric field, E, and the polarization field, P, as follows:

$$E(x, y, t) = \sum_{n=0}^{\infty} {}_n a_n(t) u_n(x, y) \tag{6.17}$$

$$P(x, y, t) = \sum_{n=0}^{\infty} {}_n b_n(t) u_n(x, y) \tag{6.18}$$

Assuming ε is constant, the spatial variation of the electric field can be expressed as follows:

$$\frac{\partial^2}{\partial t^2} E + \frac{1}{\tau_p} \frac{\partial}{\partial t} E + \omega^2 E = -\frac{1}{\varepsilon} \frac{\partial^2}{\partial t^2} P \tag{6.19}$$

where $1/\tau_p$ is used instead of σ/ε. Substitution of Eqs. (6.17) and (6.18) into (6.19) gives:

$$\sum_n \left[\ddot{a}_n u_n(x, y) + \frac{1}{\tau_p} \dot{a}_n u_n(x, y) + a_n \omega^2 u_n(x, y) \right]$$

$$= \sum_n \left(-\frac{1}{\varepsilon} \right) \ddot{b}_n u_n(x, y) \tag{6.20}$$

By multiplying both sides by $u_m^*(r)$ and integrating, the eigenfunction normalization relationship [Eq. (6.15)] can be applied, and Eq. (6.20) reduces to:

$$\ddot{a}_m + \frac{1}{\tau_p} \dot{a}_m + \omega^2 a_m = -\frac{1}{\varepsilon} \ddot{b}_m \tag{6.21}$$

By multiplying this equation by u_m, we can rewrite Eq. (6.21) as:

$$\ddot{E}_m + \frac{1}{\tau_p} \dot{E}_m + \omega^2 E_m = -\frac{1}{\varepsilon} \ddot{b}_m u_m(x, y) \tag{6.22}$$

In the case where there is only a small spatial discrepancy between P and E_m, we can write:

$$\ddot{E}_m + \frac{1}{\tau_p}\dot{E}_m + \omega^2 E_m = \frac{1}{\varepsilon}\zeta_m \ddot{P} \tag{6.23}$$

where ξ_m is the confinement factor which expresses the spatial contribution of the polarization to the mode m.

6.3. NORMAL MODES IN DISTRIBUTED INDEX (DI) WAVEGUIDES

In this section we will obtain expressions for normal modes in a DI waveguide (Fig. 6.3) for the case where the distribution of the refractive index varies as the square of the transverse distance. This mathematical treatment is useful for the following reasons:

1. This kind of waveguide is sometimes utilized in semiconductor lasers.
2. It is possible to derive a free space beam mode by extending the analytical method employed.
3. This approach can be used to derive the Fresnel–Kirchhoff (FK) integral.

The beam mode and FK integral in free space can be derived by solving the

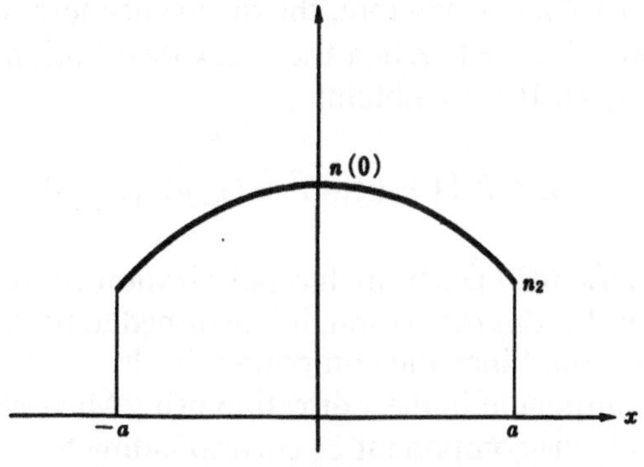

Figure 6.3. Distributed index waveguide.

Helmholtz equation [Eq. (6.10)]. We begin by assuming that the refractive index of a waveguide can be expressed as:

$$n^2(x,y) = n^2(0)[1-(g_1x)^2-(g_2y)^2] \quad \begin{Bmatrix} |x| \leqq a \\ |y| \leqq b \end{Bmatrix}$$
$$= n_2^2 \quad \begin{Bmatrix} |x| > a \\ |y| > b \end{Bmatrix} \tag{6.24}$$

where g_1 and g_2 are parameters of the index gradient called focusing constants, $n(0)$ is the refractive index at the center of the guide, and n_2 is the refractive index in the cladding.

This spatially varying refractive index contributes to the term containing the gradient of the dielectric constant in Eq. (6.9), which is approximated as:

$$\left|\nabla\left(\frac{\nabla\varepsilon}{\varepsilon}\cdot E\right)\right| \cong g_1^2(\text{or } g_2^2)E_t \tag{6.24a}$$

where E_t is the transverse component of the electric field. Neglecting losses in the medium ($\sigma = 0$), the time-varying term in Eq. (6.9) is written as:

$$\omega^2\mu_0\varepsilon E_t \cong k^2E_t, \qquad k = k_0n(0) = \omega\sqrt{\mu_0\varepsilon_0}n\ (0) \tag{6.24b}$$

The ratio between this term and the gradient term is approximately g_1^2/k^2. In the case where the width of the waveguide is many wavelengths and the variation of the refractive index is not so large, the value of g_1^2/k^2 may be as small as 10^{-6} to 10^{-7} and, therefore, the divergence term can be neglected and Eq. (6.10) may be used. When the index distribution of Eq. (6.24) is substituted into Eq. (6.10), we obtain:

$$\nabla^2E + k^2[1-(g_1x)^2-(g_2y)^2]E = 0 \tag{6.25}$$

Although the electric field generally has polarization components in the x, y, and z directions, for this discussion, it is assumed to be dominantly polarized in the x direction. Since the component in the z direction can be obtained from the component in the x direction using Maxwell's equations, let us now discuss only the component E_x corresponding to polarization in the x direction:

$$\nabla^2 E_x + k^2[1 - (g_1 x)^2 - (g_2 y)^2]E_x = 0 \tag{6.26}$$

We consider the sinusoidal wave $e^{-j\beta z}$ traveling in the z direction. Assuming no z variation of the amplitude, the component E_x may be written as:

$$E_x = \phi(x, y)e^{-j\beta z} \tag{6.27}$$

where $\phi(x, y)$ expresses the lateral distribution. Then, from Eq. (6.26) we obtain:

$$H_0\phi = \beta^2\phi \tag{6.28}$$

where

$$H_0 = \frac{\partial^2}{\partial x^2} + \frac{\partial^2}{\partial y^2} + k^2[1 - (g_1 x)^2 - (g_2 y)^2] \tag{6.29}$$

This is the eigenvalue equation associated with the operator H_0, and β^2 can be obtained as its eigenvalue. If the function ϕ is assumed to be separable into $\phi(x, y) = X(x)\cdot Y(y)$, Eq. (6.28) can be written as:

$$\frac{1}{X}\frac{d^2X}{dx^2} - k^2(g_1 x)^2 + \frac{1}{Y}\frac{d^2Y}{dy^2} - k^2(g_2 y)^2 = \beta^2 - k^2 \tag{6.30}$$

In order that the right-hand side becomes the constant $\beta^2 - k^2$ regardless of the value of the variables on the left-hand side, each expression for x and y must lead to a constant, and the following equations are obtained:

$$\frac{1}{X}\frac{d^2X}{dx^2} - k^2 g_1^2 x^2 = -\kappa_1^2 \tag{6.31a}$$

$$\frac{1}{Y}\frac{d^2Y}{dy^2} - k^2 g_2^2 y^2 = -\kappa_2^2 \tag{6.31b}$$

$$\beta^2 = k^2 - \kappa_1^2 - \kappa_2^2 \tag{6.32}$$

Equation (6.31a) is rewritten as:

$$\frac{d^2X}{dx^2} + (\kappa_1^2 - k^2 g_1^2 x^2)X = 0 \tag{6.33}$$

Here, we define

$$w_{01} = 1/\sqrt{kg_1} \tag{6.34}$$

By changing the variable to $\xi = x/w_{01}$, Eq. (6.33) becomes

$$\frac{d^2X}{d\xi^2} + (\kappa_1^2 w_{01}^2 - \xi^2)X = 0 \tag{6.35}$$

In the above equation when

$$\frac{d^2X}{d\xi^2} + (2p + 1 - \xi^2)X = 0 \tag{6.35a}$$

$X(\xi)$ can be expressed in a finite series.

The solution of the equation is given as[1]:

$$X(\xi) = N_p H_p(\xi) e^{-1/2\xi^2} \tag{6.36}$$

where $H_p(\xi)$ is the Hermite polynomial of pth order and expressed as:

p	$H_p(\xi)$
0	1
1	2ξ
2	$4\xi^2 - 2$
3	$8\xi^3 - 12\xi$
⋮	

(6.37)

The solutions, $X(\xi)$, then constitute a complete set of eigenfunctions with eigenvalues, $k_1^2 w_{01}^2 = 2p + 1$. From Eq. (6.34), κ_1^2 may be written as:

$$\kappa_1^2 = kg_1(2p + 1) \tag{6.38}$$

To normalize $X(\xi)$, it must satisfy the following equation:

$$\int_{-\infty}^{\infty} X^2(\xi)d\xi = 1 \tag{6.39}$$

from which we obtain the normalization constant, which is written as:

$$N_p = \frac{1}{[2^p p! \sqrt{\pi \omega_{01}}]} \tag{6.40}$$

The final solution to Eq. (6.35) may be written as normalized eigenfunctions as follows:

$$u_p(x, w_{01}) = \frac{1}{[2^p p! \sqrt{\pi \omega_{01}}]^{1/2}} Hp\left(\frac{x}{\omega_{01}}\right) \cdot \exp[-1/2(x/\omega_{01})^2] \tag{6.41}$$

The parameter, w_{01}, which is defined in Eq. (6.34), is a constant called the spotsize. The mode function of Eq. (6.41) is the product of a Hermite polynomial and the Gaussian function and is called a Hermite–Gaussian mode function. Figure 6.4 shows the forms of these Hermite–Gaussian functions for $p = 0, 1, 2$, and 3.

The eigenfunctions in the y direction are derived similarly, leading to:

$$Y(y) = u_q(y, w_{02}) \tag{6.42}$$

$$\kappa_2^2 = kg_2(2q + 1) \tag{6.43}$$

$$w_{02} = 1/\sqrt{kg_2} \tag{6.44}$$

Consequently, the mode function, $\phi_{pq}(x, y)$, is expressed by the following equation:

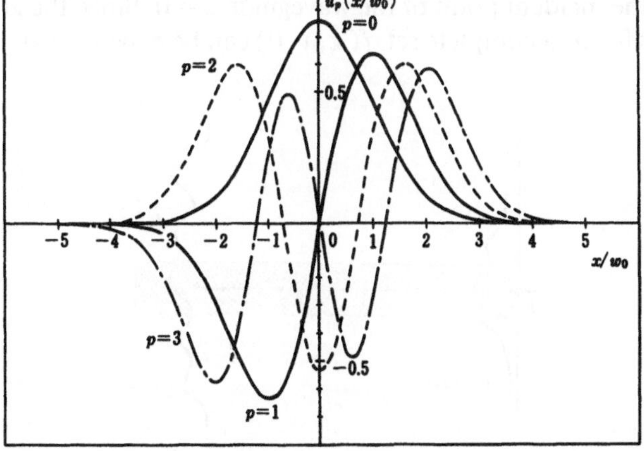

Figure 6.4. Hermite–Gaussian mode functions.

$$\phi_{pq}(x, y) = E_0 u_p(x, \omega_{01}) u_q(y, \omega_{02}) \cdot \exp[-j\beta_{pq} z] \tag{6.45}$$

where E_0 is the constant denoting the amplitude of the electric field. The characteristic propagation constant is:

$$\beta_{pq} = \sqrt{k^2 - kg_1(2p+1) - kg_2(2q+1)} \tag{6.46}$$

In the range where

$$k \gg g_1(2p+1), g_2(2q+1) \tag{6.47}$$

β_{pq} can be approximated as

$$\beta_{pq} \cong k - (p + 1/2)g_1 - (q + 1/2)g_2 \tag{6.48}$$

6.4. EXPANSION METHODS FOR NORMAL MODES

In the previous section the normal modes of the DI waveguide (specifically, a multimode waveguide with a parabolic index-of-refraction distribution) were discussed. When one of the normal modes is excited at the entrance to the waveguide, it propagates with the required propagation constant without changing its form. Therefore, if the incident beam has an arbitrary distribution, its power is partitioned into numerous propagation modes, each of which transfers the field at its own group velocity.

As is shown in Fig. 6.5, let us assume that the electric field, $f(x, y, 0)$, is excited at the incident point of the waveguide $z = 0$. Since the group of normal modes forms a complete set, $f(x, y, 0)$ can be expanded as:

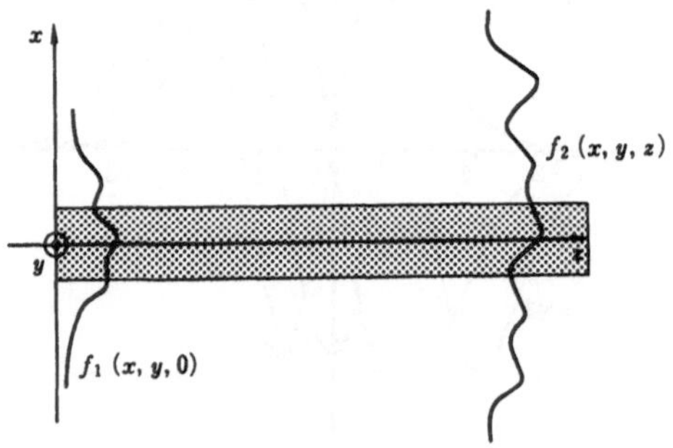

Figure 6.5. Image transformation by square-distributed index waveguide.

$$f_1(x, y, 0) = \sum_{p=0,q=0}^{\infty} a_{pq} u_p(x, \omega_{01}) u_q(y, \omega_{02}) \tag{6.49}$$

By multiplying both sides of the equation by $u_{p'}(x, w_{01}) \cdot u_{q'}(y, w_{02})$, and integrating, we find the expression for the constant a_{pq}:

$$a_{p'q'} = \int_{-\infty}^{\infty} \int_{-\infty}^{\infty} f_1(x, y, 0) u_{p'}(x, \omega_{01}) \times u_{q'}(y, \omega_{02}) dxdy \tag{6.50}$$

where the orthogonality of the eigenfunction has been used as follows:

$$\int_{-\infty}^{\infty} u_p(x, \omega_{01}) u_{p'}(x, \omega_{01}) dx = \delta_{pp'} \tag{6.51}$$

Each mode thus excited propagates with its characteristic propagation constant as given in Eq. (6.48) and after it propagates a distance, z, we obtain:

$$f_2(x, y, z) = \sum_{p=0,q=0}^{\infty} a_{pq} u_p(x, \omega_{01}) u_q(y, \omega_{02}) \cdot \exp[-j\beta_{pq} z] \tag{6.52}$$

Generally speaking, finding $f(x, y, z)$ by carrying out the summation of Eq. (6.52) is very difficult because of the slight phase difference of each mode. However, in the case where the approximations shown in Eq. (6.47) are applicable, the calculation of this sum becomes analytically possible. Substitution of Eqs. (6.48) and (6.50) into (6.52) leads to the expression:

$$f_2(x, y, z) = e^{-jkz} \sum_{p,q} \int_{-\infty}^{\infty} \int_{-\infty}^{\infty} f_1(x', y', 0) \cdot u_p(x', \omega_{01}) u_q(y', \omega_{02}) dx'dy'$$
$$\times u_p(x, \omega_{01}) u_q(y, \omega_{02}) \times \exp[jg_1(p + 1/2)z + jg_2(q + 1/2)z] \tag{6.53}$$

When this equation is rearranged by changing the order of the integral and the sum, it can be written:

$$f_2(x, y, z) = \frac{j}{\lambda z}\sqrt{\frac{g_1 z}{\sin g_1}\frac{g_2 z}{\sin g_2 z}}\, e^{-jkz} \times \iint dx'dy' f_1(x', y', 0)\cdot K(x, y; x', y') \quad (6.54)$$

where

$$K(x, x'; y, y') = \exp\left[-\frac{j}{2\omega_{01}^2}\cot g_1 z(x^2 - 2xx' \sec g_1 z + x'^2)\right] \times \exp\left[-\frac{j}{2\omega_{02}^2}\cot g_2 z(y^2 - 2yy' \sec g_2 z + y'^2)\right] \quad (6.55)$$

[The derivation of Eq. (6.54) from (6.53) is assigned later in Problem 6.3.]

Equation (6.55) has been derived by using the following Mehler's formula:

$$\sum_{n=0}^{\infty} \frac{(\frac{\xi}{2})^n}{n!} H_n(x)H_n(x') = (1 - \xi^2)^{-1/2} \exp\left[\frac{2xx'\xi - (x^2 + x'^2)\xi^2}{1 - \xi^2}\right] \quad (6.56)$$

In the case where $g_1 = g_2$, Eq. (6.55) becomes symmetrical about the axis, and, therefore, it is convenient to use cylindrical coordinates.

A similar result can be derived by using a normal mode expansion of Laguerre–Gaussian modes which are derived directly from the wave equations. Instead of this approach, however, we reach the same result by applying the following cylindrical transformation to Eq. (6.55):

$$x = r\cos\theta$$

$$y = r\sin\theta$$

By direct substitution we obtain:

$$f_2(r, \theta, z) = \frac{j}{\lambda z}\cdot\frac{gz}{\sin gz}\cdot e^{-jkz} \iint r'dr'\cdot\theta' f_1(r', \theta', 0)K(r, \theta; r', \theta') \quad (6.57a)$$

where

$$K(r, \theta; r', \theta') = \exp\left[-\frac{j}{2\omega_0^2}\cot gz \times \{r^2 - 2rr'\cos(\theta-\theta')\sec gz + r'^2\}\right] \quad (6.57b)$$

Here, we have let $w_{01} = w_{02} - w_0$ and have defined

$$w_0^2 = 1/gk \quad (6.58)$$

The integral expressions in Eqs. (6.54) and (6.57a) can, in general, be written as:

$$f(x, y, z) = \int S \int f_1(x', y', 0) G(x, y; x', y') dx'dy' \quad (6.59)$$

where $G(x, y; x', y')$ is a Green's function denoting a spatial impulse response. This integral transformation can be viewed as an extended Fresnel-Kirchhoff integral denoting the response of the waveguide to an arbitrary image $f(x, y, 0)$.

When the guiding power of the waveguide becomes weak, its properties become similar to those of uniform medium, such as free space. If we let the focusing constant, g, go to zero, Eq. (6.54) becomes:

$$f_2(x, y, z) = \frac{j}{\lambda z} e^{-jkz} \iint dx'dy' f_1(x', y', 0) \times \exp\left[-\frac{jk}{2z}\{(x-x')^2 + (y-y')^2\}\right] \quad (6.60)$$

or, in cylindrical coordinates, Eq. (6.57) becomes:

$$f_2(r, \theta, z) = \frac{j}{\lambda z} e^{-jkz} \iint r'dr'd\theta f_1(r', \theta, 0) \times \exp\left[-\frac{jk}{2z}\{r'^2 - 2rr'\cos(\theta-\theta') + r^2\}\right] \quad (6.61)$$

The above equations are the well-known Fresnel–Kirchhoff integral expressed in the paraxial approximation and mathematically represent Huygens's principle of wave propagation.

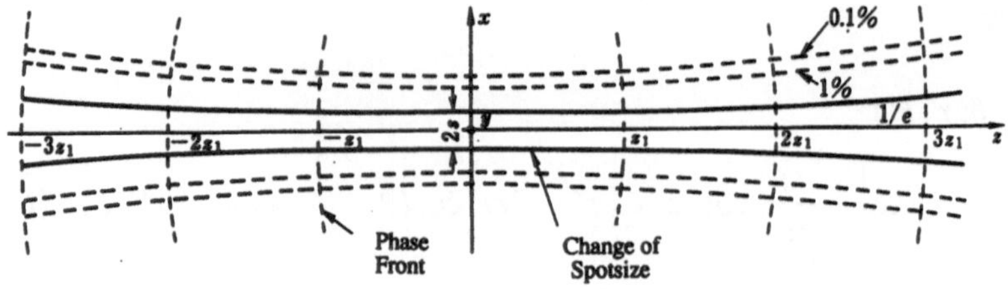

Figure 6.6. Propagation of Gaussian beams. Spotsize $s = 0.5$ mm at $z = 0$. The spotsize change is associated with $z_1 = 1$ m in this figure. The radius of curvature is enlarged for emphasis.

6.5. GAUSSIAN BEAMS IN FREE SPACE

In the previous section, the Fresnel–Kirchhoff integral was derived. We are now ready to calculate the propagation of an arbitrary incident wave so that we may determine its field and phase distribution as a function of distance. First, the propagation of beams with Gaussian distributions is considered (Fig. 6.6). The initial electric field distribution of a Gaussian beam[2–16] having a spotsize,* s, is taken to be:

$$f_1(x, y, 0) = E_0 \exp[-\tfrac{1}{2}(x^2 + y^2)/s^2] \qquad (6.62)$$

Substituting Eq. (6.62) into (6.60) leads to:

$$f_2(x, y, z) = E_0 \frac{j}{\lambda z} e^{-jkz} \int_{-\infty}^{\infty} dx' \exp\left[-\frac{1}{2}\left(\frac{x'}{s}\right)^2 - j\frac{k}{2z}(x'-x)^2\right] \quad \text{(same in } y\text{)} \qquad (6.63)$$

Here, Eq. (6.63) is a combination of a Gauss integral and a Fresnel integral.

By complex integration, we find:

$$f_2(x, y, z) = E_0 e^{-jkz}(s/w) \times \exp[-\tfrac{1}{2}P(x^2 + y^2) + j\phi] \qquad (6.64)$$

where

$$w = s\sqrt{1 + (z/ks^2)^2} \qquad \text{(spotsize)} \qquad (6.65)$$

$$R = z[1 + (ks^2/z)^2] \qquad \text{(radius of wave front)} \qquad (6.66)$$

and the parameters P and ϕ are defined as:

* Sometimes, the spotsize is defined as the e^{-1} field radius w'. In this case, $w' = 2\sqrt{w}$.

$$P = \frac{1}{w^2} + j\frac{k}{R} \qquad \text{(wave coefficient)} \tag{6.67}$$

$$\phi = \tan^{-1}(z/ks^2) \qquad \text{(phase shift)} \tag{6.68}$$

From Eq. (6.64), it can be seen that the field amplitude distribution remains Gaussian, although its spotsize and equi-phase front change. The parameter P contains an imaginary term, $j(k/R)$. By combining this imaginary term and $\exp(-jkz)$ in Eq. (6.64), we can find the equi-phase condition:

$$kz + (k/2R)r^2 = \text{constant} \tag{6.69}$$

From this equation, the functional dependence of the phase front, z, becomes:

$$z = -\frac{1}{2R}r^2 \tag{6.70}$$

This expression represents a paraboloid with a radius of curvature R. When R is positive, the phase front is convex, as seen from $z = +\infty$.

Let us also examine the parameter z/ks^2 that appears in Eq. (6.65). Recalling that $k = 2\pi/\lambda$, this parameter may be rewritten as:

$$\frac{z}{ks^2} = \frac{1}{2\pi}\cdot\frac{\lambda z}{s^2} = \frac{1}{2\pi}\left(\frac{s^2}{\lambda z}\right)^{-1} \tag{6.71}$$

which leads to the Fresnel number N which is defined as:

$$N = \frac{s^2}{\lambda z} \tag{6.72}$$

The Fresnel number is a function of the wavelength, λ, the distance, z, and the spotsize, s. It expresses a normalized inverse distance. Diffraction can be categorized in terms of N as follows:

$N \ll 1$ (Fraunhofer diffraction—far field)

$N \geq 1$ (Fresnel diffraction—near field)

When the point of observation is located far from the origin ($N \ll 1$), spotsize w can be approximated from Eq. (6.65) as:

$$w \cong \frac{z}{ks} \tag{6.73}$$

The spreading angle, $\Delta\theta$, corresponding to that spotsize is, therefore:

$$2\Delta\theta = \frac{2w}{z} = \frac{2}{ks} = \frac{2}{\pi}\cdot\frac{\lambda}{2s} = 0.64\cdot\frac{\lambda}{2s} \tag{6.74}$$

This is analogous to the spreading angle of the main lobe of a diffracted wave from a circular aperture which is given by:

$$2\Delta\theta \cong 1.22\frac{\lambda}{D} \tag{6.75}$$

Equation (6.73) can be rewritten as:

$$\frac{w}{z}\cdot k\cdot s = 1 \tag{6.76}$$

The term $(w/z)k$ can be viewed as an uncertainty in the propagation vector k_x. Since the momentum $p_x = \hbar k_x$, this looks like an uncertainty in momentum, Δp_x. The initial spotsize, s, is equivalent to an uncertainty in the x direction, Δx. Therefore, we can write:

$$\Delta p_x \times \Delta x = \hbar \tag{6.77}$$

This is just an expression of the uncertainty principle which states that, when the size of the beam is Δx at the incident point, the uncertainty Δp_x of the transverse momentum caused by the diffraction of the light is just equal to a finite value $\hbar$ ($=h/2\pi$).

We have assumed at the beginning of the section that the phase front of the incident Gaussian beam is a plane wave. One might ask if the complex number expressing the phase term in Eq. (6.62) has no imaginary part, then how does the Gaussian beam generate a curvature as the wave front propagates? In order to find the answer to this question, it is convenient to consider the conjugate propagating wave. Taking the complex conjugate of the propagating wave given in Eq. (6.64) is equivalent to replacing z by $-z$, as shown in Fig. 6.6, that is, the complex conjugate is the wave which has the same focal spot as Eq. (6.62), but propagates in the opposite direction. Thus, among all possible Gaussian propagating waves with arbitrary spotsizes and wave-front curvatures, there always exists a uniquely determined Gaussian beam which has a plane wave front and a beam waist at the locations $z = 0$, as expressed in Eq. (6.62). Bearing this fact in mind, we can obtain the form of the propagating Gaussian beam shown in Eq. (6.64) without having to calculate the complex integral, and this approach is discussed in the next section.

6.6. TRANSFORMATION MATRIX OF WAVEFORM AND RAY TRANSFER MATRIX

6.6.1. Transformation Matrix of Waveforms

From Eq. (6.64) in the previous section, we have seen that the propagating Gaussian beam can be uniquely described by the parameter P. As shown in Eq. (6.67), the complex parameter P represents both a Gaussian beam spotsize and a wave-front curvature. Therefore, we call P a waveform coefficient.[9–10] Figure 6.7 shows the waveform coefficients P_0, P_1, and P_2 at $z = 0$, $z = z_1$, and $z = z_2$, respectively. If the spotsizes and radii of curvature at each of these locations are given by s, w_1 and w_2, and R_1 and R_2, respectively, then the waveform coefficients can be expressed as:

$$P_0 = \frac{1}{s^2} \tag{6.78}$$

$$P_1 = \frac{1}{w_1^2} + j\frac{k}{R_1} \tag{6.79}$$

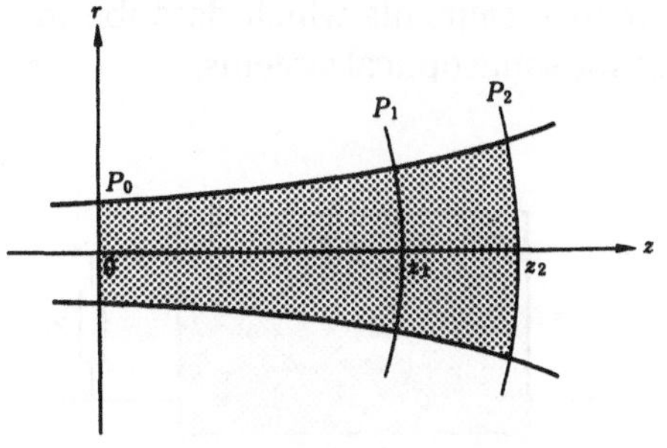

Figure 6.7. Transformation of waveform coefficients.

$$P_2 = \frac{1}{w_2^2} + j\frac{k}{R_2} \tag{6.80}$$

From Eqs. (6.78) through (6.80), we can derive the expressions:

$$\frac{1}{P_0} = \frac{1}{P_1} + j\frac{z_1}{k} \tag{6.81}$$

$$\frac{1}{P_0} = \frac{1}{P_2} + j\frac{z_2}{k} \tag{6.82}$$

When P_0 is eliminated from these two expressions, the relationship between P_1 and P_2 is reduced to:

$$P_1 = \frac{P_2}{1 + j\frac{1}{k}(z_2 - z_1)P_2} \tag{6.83}$$

This is a special case of the linear transformation:

$$P_1 = \frac{AP_2 + B}{CP_2 + D} \tag{6.84}$$

By way of analogy, a linear circuit of Fig. 6.8 has the following relationship between the impedances Z_1 and Z_2:

$$Z_1 = \frac{AZ_2 + B}{CZ_2 + D} \tag{6.85}$$

where A, B, C, and D are F matrix elements. Waveform coefficients of Gaussian beams are given by a similar transformation. The following examples will show the matrix elements which describe the transformation of Gaussian waveforms for some optical systems.

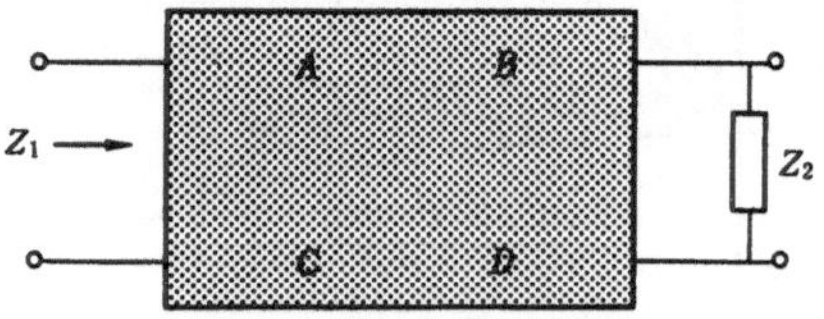

Figure 6.8. Linear circuit.

1. *Free Space*
From Eq. (6.83):

$$\tilde{F} = \begin{bmatrix} A & B \\ C & D \end{bmatrix} = \begin{bmatrix} 1 & 0 \\ j\dfrac{z}{k} & 1 \end{bmatrix} \tag{6.86}$$

where $z = z_2 - z_1$ and represents an arbitrary propagation distance in free space.

2. *Thin Convex Lens*
Figure 6.9 shows a thin convex lens with a central thickness d, radii of curvature R_1 and R_2 at the left and right sides, respectively, and refractive index, n. Because the lens is thin, we assume that there is no change in the spotsize of the Gaussian beam as it passes through the lens. Only a phase difference is added. This is called the thin lens approximation. In that case, the phase transformation may be written as:

$$e^{-j\phi(r)} \tag{6.87}$$

where $\phi(r)$ is a phase delay as the light passes from the input plane to the exit plane of the lens. It can be expressed as:

$$\begin{aligned}\phi(r) &= kn\left(d - \frac{1}{2R_1}r^2 - \frac{1}{2R_2}r^2\right) + k\left(\frac{1}{2R_1}r^2 + \frac{1}{2R_2}r^2\right) \\ &= knd - k(n-1)\left(\frac{1}{2R_1} + \frac{1}{2R_2}\right)r^2\end{aligned} \tag{6.88}$$

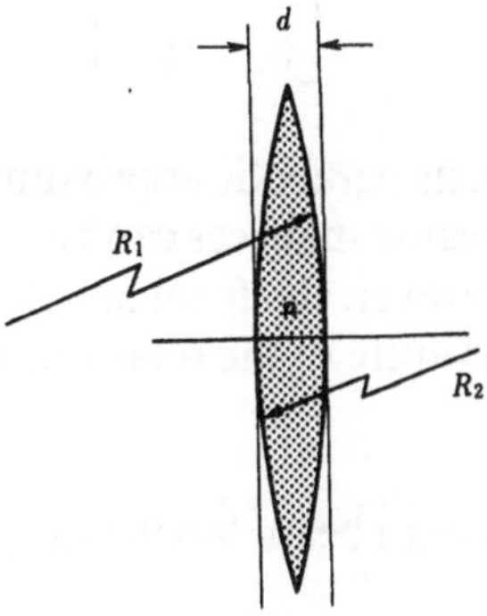

Figure 6.9. Thin convex lens with refractive index n.

$$= knd - \frac{k}{2f}r^2 \qquad (r \leqq a)$$

where a is the radius of the lens and the focal length, f, is defined as:

$$\frac{1}{f} = (n-1)\left(\frac{1}{R_1} + \frac{1}{R_2}\right) \tag{6.89}$$

Equation (6.88) shows that the lens introduces a phase delay, knd, plus a wave-front curvature, $R = -f$. Then the wave-front coefficient, P_1, of the incident wave is transformed to P_2 by the expression:

$$P_1 - j\frac{k}{f} = P_2 \tag{6.90}$$

where the constant phase change, knd, has been ignored. This then leads to the transformation matrix:

$$\tilde{F} = \begin{bmatrix} 1 & j\frac{k}{f} \\ 0 & 1 \end{bmatrix} \tag{6.91}$$

3. *Thin Concave Lens*

The transformation matrix for a thin concave lens with a focal length $-f$ can be obtained in the same way as in the case of the convex lens discussed above. This yields the expression:

$$\tilde{F} = \begin{bmatrix} 1 & -j\frac{k}{f} \\ 0 & 1 \end{bmatrix}. \tag{6.92}$$

Furthermore, using an on-axis parabolic approximation for convex and concave reflecting mirrors, the same matrices can be used as given in Eqs. (6.91) and (6.92). In this case, however, the focal length of the mirror is $f = R/2$, where R is the radius of curvature of the reflecting surface.

4. *DI Waveguides*

We assume that a Gaussian beam having the profile:

$$f_1(x, \theta, 0) = \exp[-\tfrac{1}{2} P_1 r^2] \tag{6.93}$$

is incident at the input end of a DI waveguide of length z. The output can be obtained by carrying out the integral given in Eq. (6.57a). The derivation of the transfer matrix is described in the Appendix of this chapter and leads to the expression:

$$\tilde{F} = \begin{bmatrix} \cos gz & jkg \sin gz \\ j\dfrac{1}{kg}\sin gz & \cos gz \end{bmatrix} \tag{6.94}$$

In the case of a negative DI waveguide lens in which the refractive index distribution is small at the center and becomes large at the outer edge, the matrix is given by replacing g in the above expression by jg:

$$\tilde{F} = \begin{bmatrix} \cosh(gz) & -jkg \sinh(gz) \\ j\dfrac{1}{kg}\sinh(gz) & \cosh(gz) \end{bmatrix}$$

When the optical system is composed of multiple tandem components which have transformation matrices given by $F_1, F_2, \ldots, F_n$, where these matrices describe the sequence of optical elements as the light propagates, the total matrix, F, is expressed as the product of these matrices.

$$\tilde{F} = \tilde{F}_1\tilde{F}_2 \cdots \tilde{F}_n \tag{6.95}$$

This is a general characteristic of linear transformations (see Problem 6.13).

6.6.2. Ray Transfer Matrix

Figure 6.10 shows that the ray position, x_1, and the ray slope, $\dot{x}_1$, at the incident position of an optical element are related to x_2 and $\dot{x}_2$ by the equation:

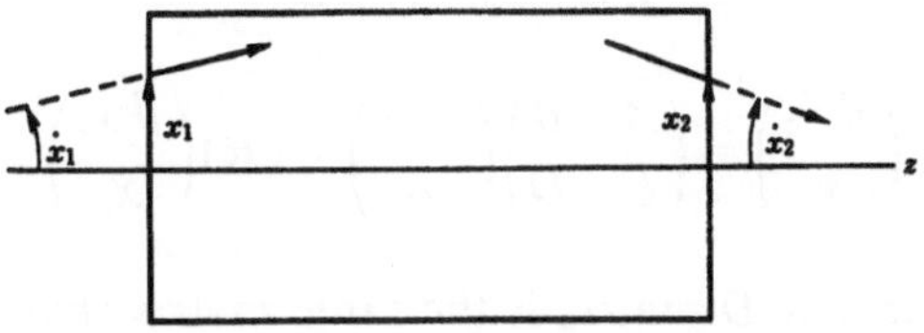

Figure 6.10. Transformation of rays.

Table 6.2. $\tilde{F}$ Matrix and Ray Matrix

Optical system	Waveform matrix, $\tilde{F}$, and ray matrix, $\tilde{R}$
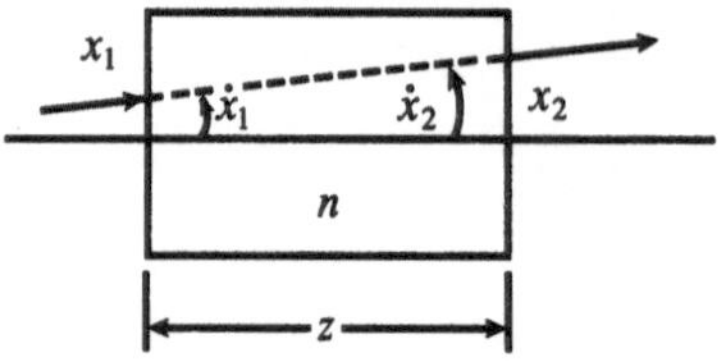 Free space	$\begin{pmatrix} 1 & 0 \\ j\frac{z}{k} & 1 \end{pmatrix}$ $k = k_0 n$ (n, refractive index)
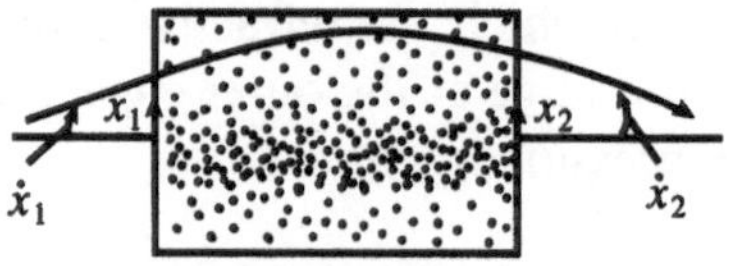 Distributed-index lens (positive)	$\begin{pmatrix} \cos gz & jkg \sin gz \\ j\frac{1}{kg} \sin gz & \cos gz \end{pmatrix}$ $k = k_0 n\ (0)$
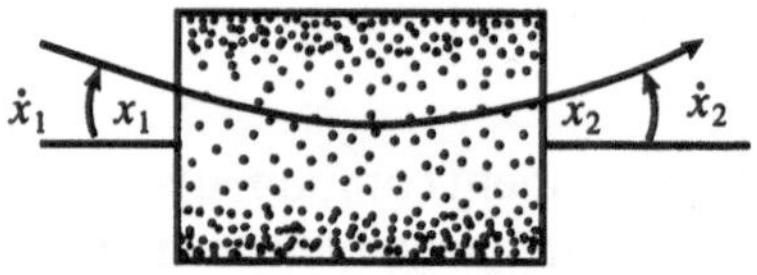 Distributed-index lens (negative)	$\begin{pmatrix} \cosh gz & -jkg \sinh gz \\ j\frac{1}{kg} \sinh gz & \cosh gz \end{pmatrix}$ $k = k_0 n\ (0)$
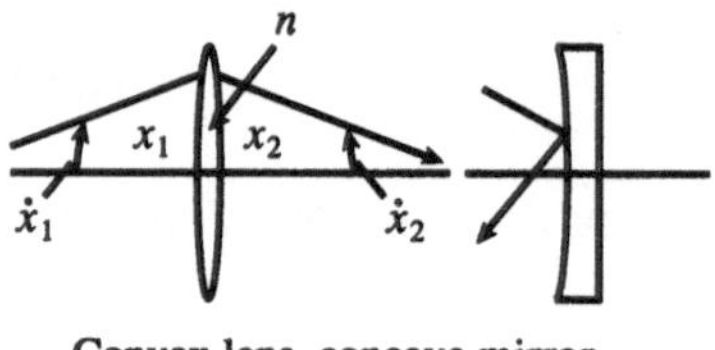 Convex lens, concave mirror	$\begin{pmatrix} 1 & \frac{jk}{f} \\ 0 & 1 \end{pmatrix}$ $k = k_0 n$ (f, focal length)
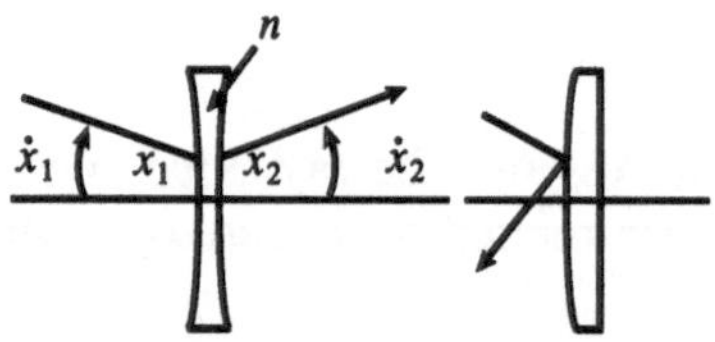 Concave lens, convex mirror	$\begin{pmatrix} 1 & \frac{jk}{f} \\ 0 & 1 \end{pmatrix}$ $k = k_0 n$ (f, virtual focal length)

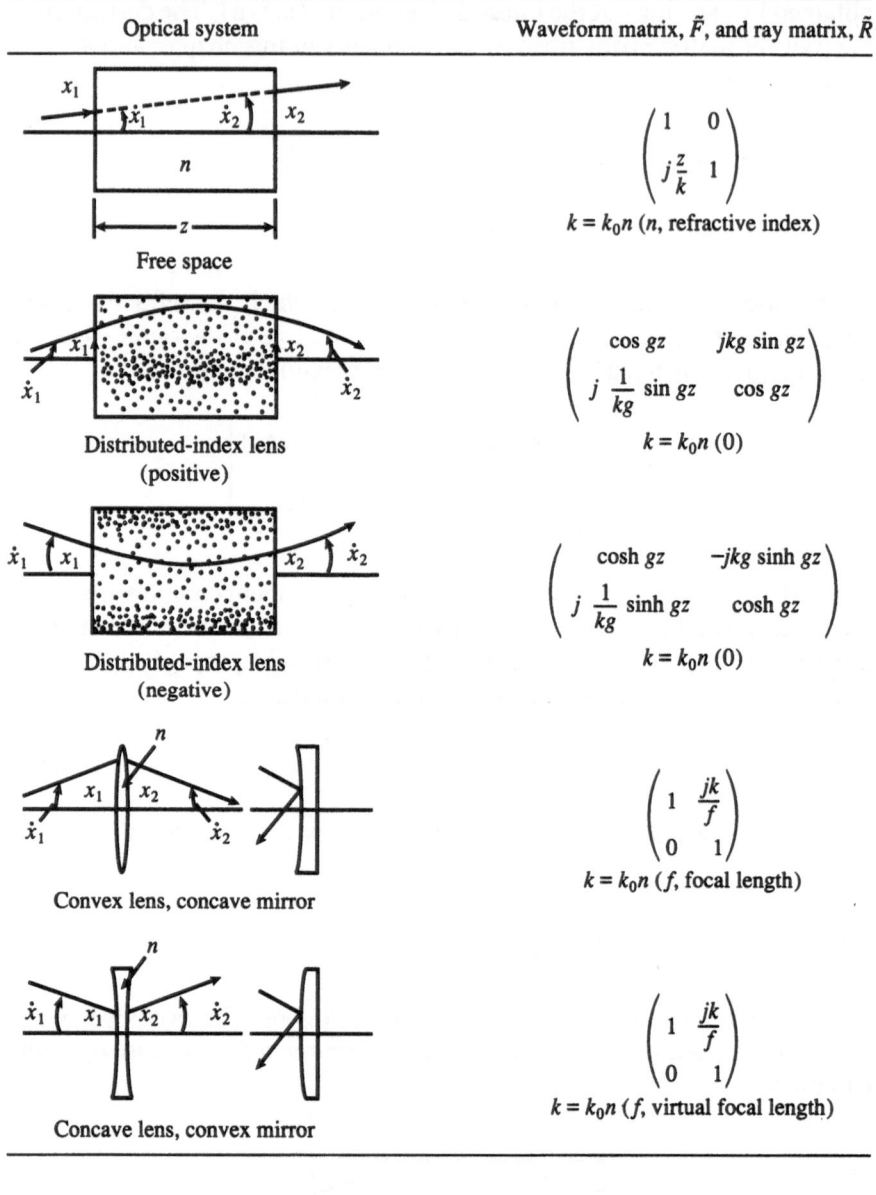

$$\begin{pmatrix} jk\dot{x}_1 \\ x_1 \end{pmatrix} = \begin{pmatrix} A & B \\ C & D \end{pmatrix} \begin{pmatrix} jk\dot{x}_2 \\ x_2 \end{pmatrix} \equiv (\tilde{R}) \begin{pmatrix} jk\dot{x}_2 \\ x_2 \end{pmatrix} \tag{6.96}$$

The ray transfer or *ABCD* matrix is the same matrix that was discussed in the previous section (see Table 6.2).

The derivation of Eq. (6.96) is easy in the case of thin lenses and reflecting mirrors. However, in the case of DI waveguides, the transmission of an off-axis incidence Gaussian beam can be solved with the help of Eq. (6.54) by solving a ray equation.[17,18]

6.7. REPRESENTATION OF WAVEFORM COEFFICIENT TRANSFORMATION BY THE SMITH CHART

The transformation of waveform coefficients through an optical component can be carried out easily on a Smith chart because it is a linear transformation with complex variables. In order to normalize the parameters, we divide both sides of Eqs. (6.79) and (6.80) by the propagation constant and obtain:

$$\begin{aligned} P_1/k &= \frac{1}{kw_1^2} + j\frac{1}{R_1} = u_1 + jv_1 = p_1 \\ P_2/k &= \frac{1}{kw_2^2} + j\frac{1}{R_2} = u_2 + jv_2 = p_2 \end{aligned} \tag{6.97}$$

Then the transformation relating to the optical element is reduced to:

$$u_2 + jv_2 = \frac{D(u_1 + jv_1) - B/k}{-kC(u_1 + jv_1) + A} \tag{6.98}$$

Let us consider the transformation of waveform coefficients for lenses and free space as typical examples.

1. *Transformation in Lenses*

The transformation of a thin lens having a focal length f is written from Eq. (6.90) as:

$$u_2 + jv_2 = u_1 + j(v_1 - 1/f) \tag{6.99}$$

This transformation can be carried out by shifting a distance $1/f$ along a curve with a constant real part (resistance) on the Smith chart.

2. *Transformation in Free Space*

From Eq. (6.86) the transformation in free space over a distance z is written as:

$$u_2 + jv_2 = \frac{1}{\dfrac{1}{u_1 + jv_1} - jz} \tag{6.100}$$

This transformation can be carried out on a Smith chart through the inverse operation of complex numbers.

Example: Waveform Coefficients of a Gaussian Beam Radiated from a Laser Resonator

Let us obtain, as an example, the waveform coefficients of a Gaussian beam with $\lambda = 0.6328\ \mu m$ (ruby laser) at $z = 1.4$ m radiated from a laser resonator shown in Fig. 6.11. The output mirror acts as a negative lens with a local length of -1.935 m for the output light passing through it. With the local length of the resonator mirrors each equal to 0.5 m, $f/l = 0.5/0.475$. This, together with the propagation constant, $k = 9.925 \times 10^6\ (l/m)$, gives the spotsize at the center of the laser $w_0 = 0.224$ mm. The waveform coefficient divided by k at the middle of the resonator is $p_0 = l/(kw_0^2) = 2(l/m)$. Figure 6.12 shows the Smith chart transformation of the coefficients from p_0 to p_3 at $z = 1.4$ m, where z and f^{-1} have been normalized by m and l/m, respectively.

From Fig. 6.12, $p_3 = 0.09 + j0.53(l/m)$; therefore, $P_3 = (0.89 + j5.26) \times 10^6(l/m^2)$ by multiplying the propagation constant, k. A precise calculation of P_3 yields $0.88 \times 10^6 + j5.23 \times 10^6(l/m^2)$.

6.8. APPENDIX 1: MATRIX OF A DI WAVEGUIDE

Positive Lenslike Medium

Transformation of Waveform Coefficients

Let us consider a positive lenslike medium for which the dielectric constant has a parabolic distribution as shown in Fig. 6.13a. The waveform co-

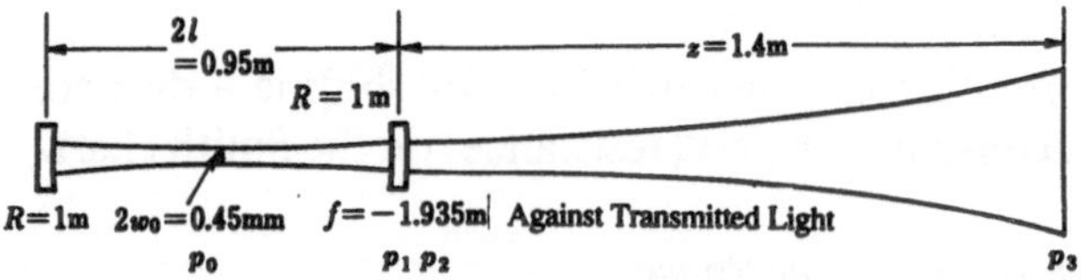

Figure 6.11. Radiation from a laser resonator ($p_1, p_2, \ldots$ are waveform coefficients normalized by k).

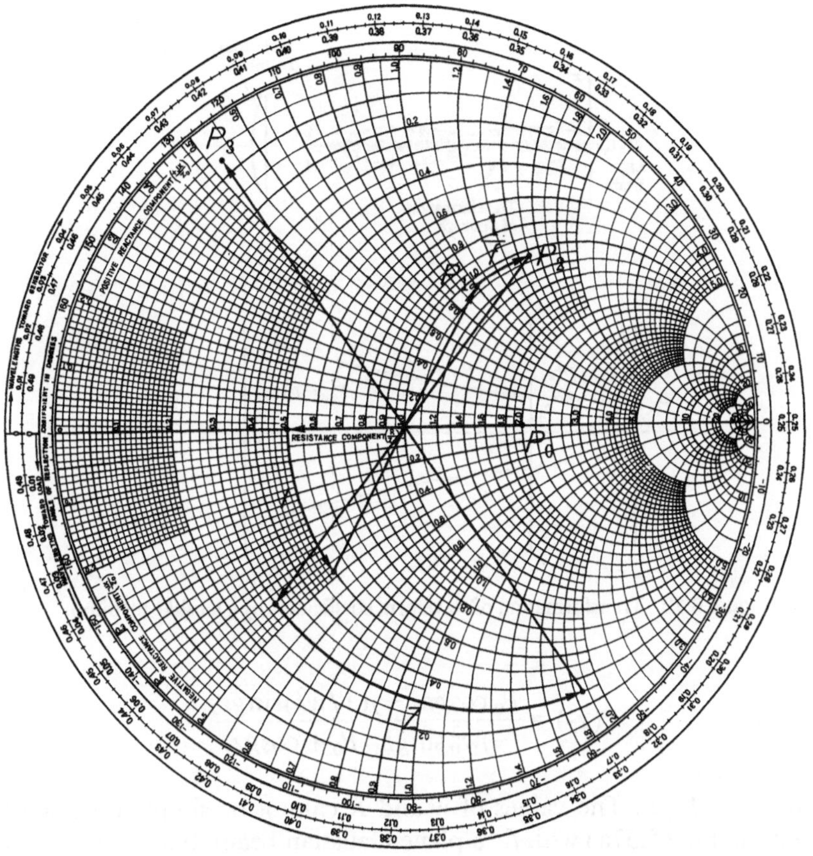

Figure 6.12. Smith chart representing transformation of waveform coefficients of Fig. 6.11. Angle of reflection coefficient in degrees: inductive reactance component (+ JX/Z$_0$) or capacitive susceptance component (+ JB/Y$_0$); capacitive reactance component (− JX/Z$_0$) or inductive susceptance component (− JB/Y$_0$).

efficients are expressed as P_{in} and P_{out}, respectively. To obtain the transformation connecting these coefficients, we consider a virtual medium of length z_1 which has the same index-of-refraction profile. Let us choose a spotsize s and a flat phase at the input side of this virtual medium. Then the waveform coefficient, P_0, is given by $P_0 = 1/s^2$.

If $P_0 = 1/s^2$, $P(z_1) = 1/w^2(z_1) + jk/R(z_1)$, and $P(z_2) = 1/w^2(z_2) + jk/R(z_2)$, then the relationship between P_0 and $P(z_1)$ and P_0 and $P(z_2)$ can be calculated from the values of w and R at z_1 and z_2 as:

$$P(z_1) = \frac{P_0 \cos gz_1 - j(1/w_0^2)\sin gz_1}{-jw_1^2 \sin gz_1 \cdot P_0 + \cos gz_1} \tag{6.101}$$

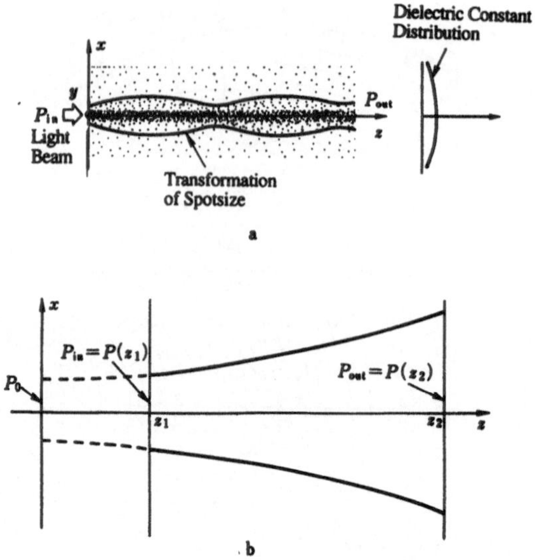

Figure 6.13. Transformation of waveform coefficients by a lenslike medium.

$$P(z_2) = \frac{P_0 \cos gz_2 - j(1/w_0^2)\sin gz_2}{-jw_0^2 \sin gz_1 \cdot P_0 + \cos gz_2} \tag{6.102}$$

where $w_0^2 = 1/gk$. The values of w and R cannot be simply obtained by integrating Eq. (6.57a) with f_1, a plane Gaussian beam, but can be derived after some lengthy calculations.

If we put $z = z_2 - z_1$ and eliminate z_0 from Eqs. (6.101) and (6.102), we obtain the expression:

$$P(z_1) = \frac{\cos gz \cdot P(z_2) + j(1/w_0^2)\sin gz}{jw_0^2 \sin gz_1 \cdot P(z_2) + \cos gz} \tag{6.103}$$

Then the F matrix representation becomes:

$$\tilde{F} = \begin{bmatrix} \cos gz & j(1/w_0^2)\sin gz \\ jw_0^2 \sin gz & \cos gz \end{bmatrix} \tag{6.104}$$

Ray Transformation

Figure 6.14 describes the passage of an optical ray through a positive lenslike medium. If $x_0 = x_{in}$, $\dot{x}_0 = \dot{x}_{in}$, $x = x_{out}$, and $\dot{x} = \dot{x}_{out}$, then the fol-

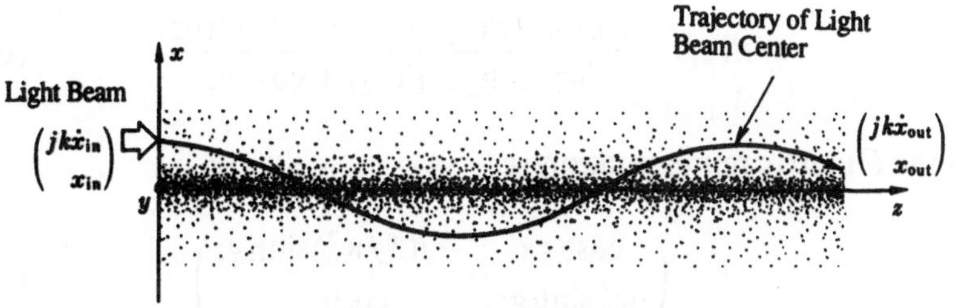

Figure 6.14. Transformation of an optical path by a positive lenslike medium.

lowing equations describe the propagation of the ray through a positive lens-like medium:

$$\left.\begin{aligned} jk\dot{x}_{\text{in}} &= \cos gz \cdot jk\dot{x}_{\text{out}} + jkg \sin gz x_{\text{out}} \\ x_{\text{in}} &= j\frac{1}{kg} \sin gz \cdot jk\dot{x}_{\text{out}} + \cos gz x_{\text{out}} \end{aligned}\right\} \tag{6.105}$$

where $w_0^2 = 1/kg$. These expressions have been written using the ray transfer matrix which is the same as the F derived in the previous section:

$$\tilde{R} = \begin{pmatrix} \cos gz & j(1/w_0^2)\sin gz \\ jw_0^2 \sin gz & \cos gz \end{pmatrix} \tag{6.106}$$

Negative Lenslike Medium

Transformation of Waveform Coefficients

Following the same approach as in the positive lenslike medium, we can obtain:

$$P(z_1) = \frac{P_0 \cosh gz_1 + j(1/w_0^2)\sinh gz_1}{-jw_0^2 \sinh gz_1 + \cosh gz_1} \tag{6.107}$$

$$P(z_2) = \frac{P_0 \cosh gz_2 + j(1/w_0^2)\sinh gz_2}{-jw_0^2 \sinh gz_2 + \cosh gz_2} \tag{6.107$'$}$$

Accordingly,

$$P(z_1) = \frac{\cosh gz \cdot P(z_2) - j(1/w_0^2)\sinh gz}{jw_0^2 \sinh gz \cdot P(z_2) + \cosh gz} \tag{6.108}$$

then the F matrix representation is:

$$\tilde{F} = \begin{pmatrix} \cosh gz & -j(1/w_0^2)\sinh gz \\ jw_0^2 \sinh gz & \cosh gz \end{pmatrix} \tag{6.109}$$

Ray Transformation

Once again, the F matrix is the same as the ray transformation matrix, so we can write for a single ray:

$$\begin{aligned} jk\dot{x}_{\text{in}} \cosh gz \cdot jk\dot{x}_{\text{out}} - jkg \sinh gz \cdot x_{\text{out}} \\ x_{\text{in}} = j(1/kg) \sinh gz \cdot jk\dot{x}_{\text{out}} + \cosh gz \cdot x_{\text{out}} \end{aligned} \tag{6.110}$$

and the ray transformation matrix is

$$\tilde{R} = \begin{bmatrix} \cosh gz & -j(1/w_0^2)\sinh gz \\ jw_0^2 \sinh gz & \cosh gz \end{bmatrix} \tag{6.111}$$

6.9. APPENDIX 2: TRANSFER MATRICES IN FREE SPACE

By letting g go to zero in Eqs. (6.104) and (6.106), the F matrix and the ray transfer matrix are given by:

$$\tilde{F} = \begin{bmatrix} 1 & 0 \\ j\dfrac{z}{k} & 1 \end{bmatrix} \tag{6.112}$$

$$\tilde{R} = \begin{bmatrix} 1 & 0 \\ j\dfrac{z}{k} & 1 \end{bmatrix} \tag{6.113}$$

These two matrices are being treated as though they are different, but they are the same matrix. It is very interesting to find that the wave approach and ray optics have brought us to the same result. This is not surprising, since both have originated from Maxwell's equation.

PROBLEMS

6.1. Explain what the completeness of the eigenfunction given by Eq. (6.16) means.
6.2. Prove that for $u_p(x, w_{01})$ in Eq. (6.41), $S_{\infty}^{b} u_p(x, \omega_{01}) u_p^1(x, \omega_{01}) \partial x = \partial_{pp'}$
6.3. Derive Eq. (6.54). This will require much analytic work, but you will be compensated when you reach the final result.
6.4. Show that Eq. (6.57) is obtained by a coordinate transformation.
6.5. Derive Eqs. (6.60) and (6.61).
6.6. Obtain Eq. (6.64).
6.7. In obtaining Eq. (6.70), we have assumed that R is constant. Consider how the result changes when taking Eq. (6.66) into consideration.
6.8. Obtain the Fresnel number N for $\lambda = 1.3\ \mu\text{m}$, $z = 10$ cm, and $s = 1$ mm.
6.9. Obtain Eq. (6.83).
6.10. Show Eq. (6.90).
6.11. Confirm Eq. (6.96).
6.12. Trace the changes of wave coefficients shown in the example of Section 6.7 on the Smith chart. Also consider the case where the wavelength $\lambda = 1.55\ \mu\text{m}$.
6.13. When two linear systems having matrices are connected in tandem

$$F_1 = \begin{bmatrix} A_1 & B_1 \\ C_1 & D_1 \end{bmatrix} \quad F_2 = \begin{bmatrix} A_2 & B_2 \\ C_2 & D_2 \end{bmatrix}$$

show that the linear transformation of total system is expressed by

$$F = F_1 \cdot F_2$$

REFERENCES

1. L. I. Schiff, *Quantum Mechanics,* 3rd ed., p. 71, McGraw–Hill, New York.
2. G. Goubau and F. Schwering, *IRE Trans. Antennas Propag.* **AP-9**(5), 248 (May 1961).
3. A. G. Fox and T. Li, *Bell Syst. Tech. J.* **40,** 453 (1961).
4. G. D. Boyd and H. Kogelnik, *Bell Syst. Tech. J.* **41,** 1347 (1962).
5. J. Hirano and Y. Fukatsu, *Proc. IEEE* **52**(11), 1284 (Nov. 1964).
6. D. W. Berreman, *Bell Syst. Tech. J.* **43**(pt. 1), 1469 (1964).
7. P. K. Tien, J. P. Gordon, and J. R. Winnery, *Proc. IEEE* **53**(2), 129 (Feb. 1965).
8. H. G. Unger, *Arch. Elekt.* Übertragung **49,** 189 (April 1965).
9. Y. Suematsu and Y. Fukinuki, *J. IECE Jpn.* **48,** 1684 (1965).
10. H. Kogelnik, *Bell Syst. Tech. J.* **44**(10), 455 (1965).
11. Y. Suematsu, K. Iga, and S. Ito, *IEEE Trans. Microwave Theory Tech.* **MTT-14**(12), 657 (Dec. 1966).
12. Y. Suematsu and K. Iga, *Trans. IECE Jpn.* **19,** 1645 (Sept. 1966).
13. H. Kogelnik and T. Li, *Appl. Opt.* **5,** 1550 (1966).
14. E. A. J. Marcatili, *Bell Syst. Tech. J.* **45,** 105 (1966).

15. S. Kawakami and J. Nishizawa, *IEEE Trans. Microwave Theory Tech.* **MTT-16**(10), 814 (Oct. 1968).
16. H. Kogelnik, in *Integrated Optics* (T. Tamir, ed.), Springer-Verlag, Berlin (1979).
17. D. Marcuse, *Light Transmission Optics,* 2nd ed., Van Nostrand–Reinhold, Princeton, N.J. (1982).
18. Y. Suematsu and K. Iga, *Fundamentals of Optical Fiber Communication,* 3rd ed., Ohmsha, Tokyo (1991).

SEVEN

OPTICAL WAVEGUIDES FOR LASER TECHNOLOGY

This chapter describes the fundamentals of dielectric optical waveguides. In the previous chapter, propagating beam modes in the distributed index (DI) waveguide and in free space were studied. In this chapter, a step index optical waveguide is discussed and the basic theory of its propagating modes is derived. Next, an equivalent refractive index method which is useful for mode analysis of three-dimensional waveguides (those which are finite in the horizontal direction as well as in the vertical direction) will be introduced. These subjects will be useful for understanding the mode in semiconductor lasers. We will also consider a mode confinement factor which directly relates to the performance of semiconductor lasers and to the radiation of light from a waveguide.

7.1. NORMAL MODES IN A PLANAR DIELECTRIC WAVEGUIDE

In this section we introduce the normal modes in a planar dielectric waveguide with a uniform index distribution in the y (horizontal) direction and a step distribution in the x (vertical) direction, as shown in Fig. 7.1a.

The physical meaning of waveguiding is understood in the following manner. If a local plane wave propagates through the cladding and core regions show in Fig. 7.1b, the propagation speed is greater in the cladding and, therefore, the propagating light refracts toward the center of the core region. However, if the light is focused to a small spot, it will spread out again by diffraction. A normal mode does not suffer any change of its wave front or intensity distribution as it propagates. Seen from another point of view, as

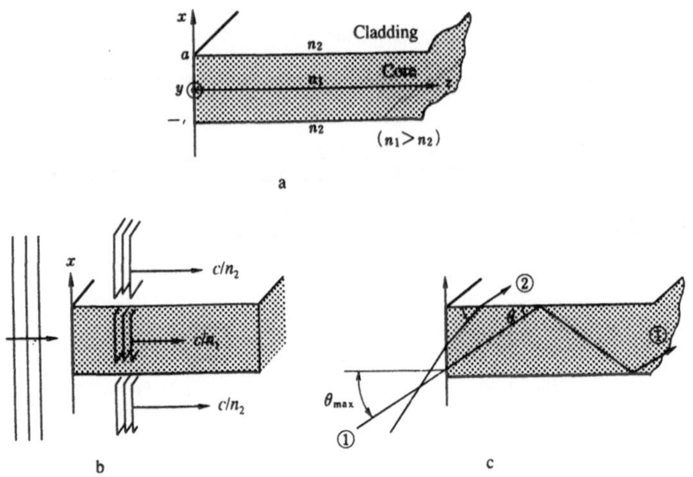

Figure 7.1. Optical waveguide. (a) Waveguide; (b) wave picture; (c) ray picture.

shown in Fig. 7.1c, the light is confined to propagate in the core by total internal reflection at the cladding boundaries. At these boundaries, the reflected wave suffers some phase shift which is associated with the excitation of an evanescent wave in the cladding material which falls off exponentially away from the boundaries. Therefore, the propagating normal mode has a well-defined distribution which extends into the cladding material.

The above-mentioned concept is a fundamental property of the waveguiding mechanism. In semiconductor lasers, waveguiding plays an important role in determining the laser performance. Figure 7.1a shows a so-called step index waveguide where the refractive index n_1 in the core is higher than n_2 in the outer part (i.e., the cladding), and the refractive index is uniform in each of these two regions. If n_1 is smaller than n_2, the waveguide is "leaky" since only some finite amount of light is reflected. But it still can guide light energy, and is called a leaky waveguide. In order to understand the principle of this type of waveguide, we shall first explain the total internal reflection of light as an example.

Among those light beams at various incident angles, θ, which are coupled into the central axis of the core, light beam 2 with a large incident angle, θ, in Fig. 7.1c passes through the core and into the cladding. Light beam 1, whose incident angle, θ, is smaller than the critical angle of total reflection, $\theta_c = 90° - \phi_c$, is totally reflected and confined within the core. If the core is sandwiched between two cladding media, then the light can propagate by repeating total internal reflections. The critical angle, θ_c, is that angle at which the light refracted into the cladding is parallel to the boundary. It can be obtained from Snell's law and is given by the expression:

$$\cos\theta_c = n_2/n_1 \tag{7.1}$$

When the index difference between n_1 and n_2 is small, θ_c becomes very small as seen from Eq. (7.1). By using $\sin^2\theta_c + \cos^2\theta_c = 1$ and $n_1^2 - n_2^2 = 2n_1(n_1 - n_2)$ when $n_1 \cong n_2$, the critical angle can be approximated as:

$$\theta_c = \sin^{-1}\sqrt{1 - \cos^2\theta_c} \simeq \sin^{-1}\sqrt{2(n_1 - n_2)/n_1} = \sin^{-1}\sqrt{2\Delta} \simeq \sqrt{2\Delta}$$

Total reflecting angle (7.2)

where

$$\Delta = (n_1^2 - n_2^2)/2n_1^2 \simeq (n_1 - n_2)/n_1 \tag{7.3}$$

Here, Δ, the normalized refractive index difference, is expressed by the ratio of the refractive index difference $(n_1 - n_2)$ to the refractive index n_1 at the core, and is an important parameter for dielectric waveguides as well as for optical fibers. Frequently, Δ is multiplied by 100 and expressed in terms of a percentage of n_1. When $\Delta = 1\%$, $\theta_c \cong \sqrt{2 \times 0.01} = 0.14$ rad $= 8°$.

The acceptance angle, $2\theta_{max}$, is defined by the maximum angle over which the waveguide can receive rays, as shown in Fig. 7.1c. Since it is measured outside of the waveguide, Snell's law gives:

$$\sin\theta_{max}/\sin\theta_c = n_1/n_0 \tag{7.4}$$

When the outside medium is air ($n_0 = 1$), we obtain:

$$2\theta_{max} = 2\sin^{-1}(n_1 \sin\theta_c) = 2\sin^{-1} n_1\sqrt{2\Delta}$$

(Maximum acceptance angle) (7.5)

The numerical aperture (NA) is given by the sin of θ_{max} and is a very important parameter indicating the light acceptance performance of the waveguide. From the above expressions the numerical aperture may be written as:

$$\mathrm{NA} = \sin\theta_{max} = n_1 \sin\theta_c = \sqrt{(n_1^2 - n_2^2)} = n_1\sqrt{2\Delta} \tag{7.5'}$$

when $n_1 = 1.5$ and $\Delta = 1\%$, NA = 0.21.

Light beams incident on the waveguide which fall within the acceptance angle propagate far without radiating from the core, though they may become weak because of absorption. As discussed later, even if the angle, θ, of the propagating light beam in the core lies within the range $0 < \theta < \theta_c$, it must have discrete values for low-loss propagation because of coupling to the normal modes mentioned later.

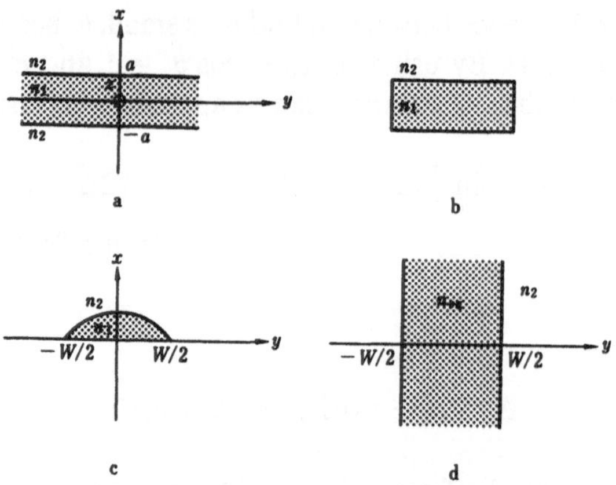

Figure 7.2. Various dielectric waveguides.

In semiconductor lasers, light is guided in this manner when the active region with a refractive index n_1 which constitutes the core is sandwiched in between the cladding layers with a refractive index $n_2 (n_1 > n_2)$, as shown in Fig. 7.2a. In a so-called mode-controlled laser where the index of refraction variation exists in the transverse direction as well, as seen in Fig. 7.2b, a three-dimensional waveguide is realized. However, since the fundamentals of waveguiding phenomena can be introduced using a two-dimensional waveguide geometry, we shall first describe the modes of such a planar two-dimensional waveguiding device.

We first introduce a guided-wave approach[1,2] to the symmetrical planar waveguide where the core, which has a refractive index n_1 and a thickness $2a$, is placed in between cladding layers which have a refractive index n_2 as shown in Fig. 7.2a. By assuming the time dependence of the electric field to be $\exp(j\omega t)$, Maxwell's equations are:

$$\Delta \times E = -j\omega\mu_0 H \tag{7.6}$$

$$\Delta \times H = +j\omega\varepsilon E \tag{7.7}$$

where μ_0 is the magnetic permeability in vacuum and ε is the dielectric constant of the medium. Here, $\partial/\partial y = 0$ because the planar waveguide is infinite in the y direction. Consequently, the propagating electromagnetic field may be grouped into two modes which have either their electric fields parallel to the y direction, or their magnetic fields parallel to the y direction:

$$\text{TE modes: } E(0, E_y, 0),\ H(H_x, 0, H_z)$$

$$\text{TM modes: } E(E_x, 0, E_z),\ H(0, H_y, 0)$$

In all cases, propagation is in the z direction. The justification for this grouping is left to the reader. We begin by deriving the propagation equation for the electric field E_y associated with the TE modes. From Eq. (7.6):

$$H_x = \frac{1}{j\omega\mu_0} \cdot \frac{\partial E_y}{\partial z}, \qquad H_z = -\frac{1}{j\omega\mu_0} \cdot \frac{\partial E_y}{\partial x} \tag{7.6'}$$

then, from Eq. (7.7):

$$\frac{\partial H_x}{\partial z} - \frac{\partial H_z}{\partial x} = j\omega\varepsilon E_y \tag{7.7'}$$

Substituting Eq. (7.6′) into (7.7′) leads to:

$$\frac{\partial^2 E_y}{\partial x^2} + \frac{\partial^2 E_y}{\partial z^2} + \omega^2\mu_0\varepsilon E_y = 0 \tag{7.8}$$

Assuming the dependence in the z direction to be $\exp(-j\beta z)$, the propagation equation for x becomes:

$$\frac{d^2 E_y}{dx^2} + (k_0^2 n^2 - \beta^2)E_y = 0 \tag{7.9}$$

Here, we have used the relation $\varepsilon = \varepsilon_0 n^2$, and a free space propagation constant:

$$k_0^2 = \omega^2\mu_0\varepsilon_0 \tag{7.10}$$

Using the index of refraction in the core or in the cladding, the following wave equations are written:

$$\frac{d^2 E_y}{dx^2} + (k_0^2 n_1^2 - \beta^2)E_y = 0 \qquad (n = n_1\text{: core}) \tag{7.11}$$

$$\frac{d^2 E_y}{dx^2} + (k_0^2 n_2^2 - \beta^2)E_y = 0 \qquad (n = n_2\text{: cladding}) \tag{7.12}$$

If we define:

$$\kappa^2 = k_0^2 n_1^2 - \beta^2 \tag{7.13}$$

and

$$\gamma^2 = \beta^2 - k_0^2 n_2^2 \qquad (7.14)$$

then the relationship between κ and γ may be expressed as:

$$\kappa^2 + \gamma^2 = k_0^2(n_1^2 - n_2^2) \qquad (7.15)$$

A normalized frequency V based on the critical angle [Eqs. (7.2) and (7.3)] is defined as:

$$V = k_0 a n_1 \sqrt{2\Delta} \qquad (7.16)$$

From the above expressions we then obtain:

$$(\kappa a)^2 + (\gamma a)^2 = V^2 \qquad (7.17)$$

Equation (7.17) is the deterministic equation (1) in Table 7.1. Since the solutions for Eq. (7.11), which describes the propagation in the core, and Eq. (7.12), which describes the propagation in the cladding, and trigonometric functions and exponential functions, respectively, it is easily understood that the eigenfunctions shown in Table 7.1 are obtained for even modes and odd modes.

From Eq. (7.6′)

$$H_z = -\frac{1}{j\omega\mu_0} \cdot \frac{\partial E_y}{\partial x} \qquad (7.18)$$

the magnetic field component, H_z, can be obtained from the electric field. For example, H_z derived from the even mode, E_y, gives:

$$H_z = A_e\left(\frac{\kappa}{j\omega\mu_0}\right)\sin\kappa x \qquad (|x| \leqq a) \qquad (7.19)$$

$$H_z = A_e \frac{x}{|x|}\cos(\kappa a)\left(\frac{\gamma}{j\omega\mu_0}\right)e^{-\gamma(|x|-a)} \qquad (|x| \geqq a) \qquad (7.20)$$

H_z is a tangential magnetic field component which must be continuous at the core–cladding boundary. From this boundary condition, Eq. (7.19) must equal Eq. (7.20) at $x = a$, and the following relations are found:

$$\kappa \sin ka = \gamma \cos\kappa a \qquad (7.21a)$$

or

Table 7.1. Mode Functions of Two-Dimensional Dielectric Waveguide

	Even mode	Odd mode	Deterministic equation
TE	$E_y = \begin{cases} A_e \cos \kappa x, & (\lvert x \rvert \leqq a) \\ A_e \cos(\kappa a) e^{-\gamma(\lvert x \rvert - a)}, & (\lvert x \rvert \geqq a) \end{cases}$	$E_y = \begin{cases} A_0 \sin \kappa x, & (\lvert x \rvert \leqq a) \\ \frac{x}{\lvert x \rvert} A_0 \sin(\kappa a) e^{-\gamma(\lvert x \rvert - a)}, & (\lvert x \rvert \geqq a) \end{cases}$	$(\kappa a)^2 + (\gamma a)^2 = V^2$ ① $\begin{cases} \tan(\kappa a) = \frac{\gamma a}{\kappa a} \text{ (even order)} & ② \\ \tan(\kappa a) = -\frac{\kappa a}{\gamma a} \text{ (odd order)} & ②' \end{cases}$
	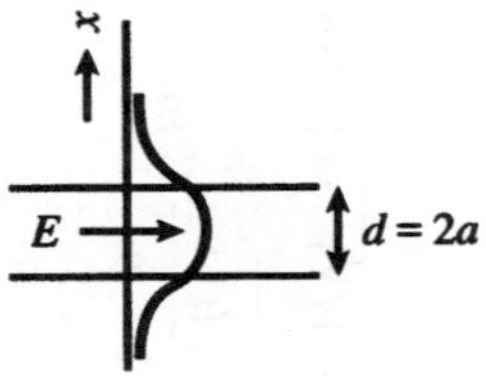 	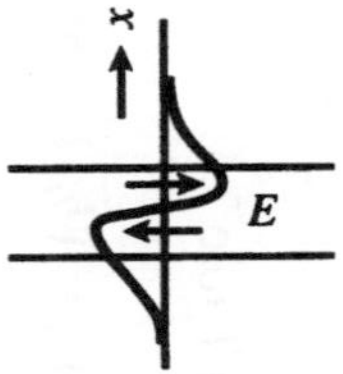 	
TM	$H_y = \begin{cases} B_e \cos \kappa x, & (\lvert x \rvert \leqq a) \\ B_e \cos(\kappa a) e^{-\gamma(\lvert x \rvert - a)}, & (\lvert x \rvert \geqq a) \end{cases}$	$H_y = \begin{cases} B_0 \sin \kappa x, & (\lvert x \rvert \leqq a) \\ \frac{x}{\lvert x \rvert} B_0 \sin(\kappa a) e^{-\gamma(\lvert x \rvert - a)}, & (\lvert x \rvert \geqq a) \end{cases}$	$(\kappa a)^2 + (\gamma a)^2 = V^2$ ① $\begin{cases} \tan(\kappa a) = \left(\frac{n_1}{n_2}\right)^2 \frac{\gamma a}{\kappa a} \text{ (even order)} & ③ \\ \tan(\kappa a) = -\left(\frac{n_2}{n_1}\right)^2 \frac{\kappa a}{\gamma a} \text{ (odd order)} & ③' \end{cases}$
	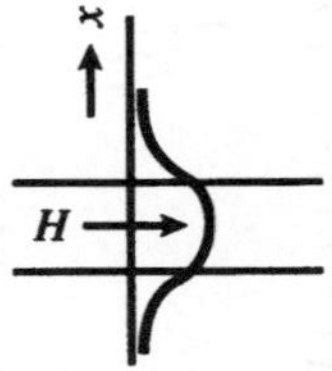 	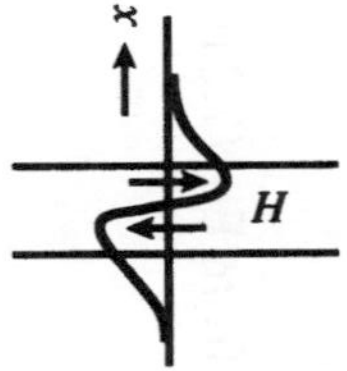 	

$$\sin\kappa a/\cos\kappa a = \gamma a/\kappa a \tag{7.21b}$$

These expressions lead to Eq. (2) in Table 7.1. For the odd modes, a similar calculation can be performed, and we find:

$$\sin\kappa a/\cos\kappa a = -\kappa\alpha/\gamma a \tag{7.21c}$$

For convenience, b is defined as a normalized propagation constant:

$$b = (\beta^2 - k_0^2 n_2^2)/(k_0^2 n_1^2 - k_0^2 n_2^2) \cong (\beta/k_0 - n_2)/(n_1 - n_2) \tag{7.22}$$

and the relationship between this constant and the normalized frequency, V, can be obtained as follows. First, from Eqs. (7.13), (7.14), and (7.22):

$$\kappa a = \sqrt{1 - b}V, \qquad \gamma a = \sqrt{b}V$$

Then, from Eq. (7.21b) or Eq. (2) in Table 7.1:

$$\tan(\sqrt{1 - b}V) = \sqrt{b/(1 - b)}$$

for even modes. Therefore, we have:

$$\therefore V = \frac{1}{\sqrt{1 - b}} \times \left[\tan^{-1}\sqrt{\frac{b}{1 - b}} + 2m\cdot\frac{\pi}{2}\right] \qquad (m = 0, 1, 2)$$

On the other hand, for the odd modes, from Eq. (7.21c) or Eq. (2′) in Table 7.1, we have:

$$\cot(\sqrt{1 - b}V) = -\sqrt{b/(1 - b)} \therefore V = 1/\sqrt{1 - b}[-\cot^{-1}(\sqrt{b/(1 - b)}) + m\pi]$$

$$= \frac{1}{\sqrt{1 - b}}\left[\tan^{-1}\sqrt{\frac{b}{1 - b}} + (2m - 1)\frac{\pi}{2}\right] \qquad (m = 1, 2, 3)$$

Consequently, we can obtain the relationship for the Nth modes[1]:

$$V = \frac{1}{\sqrt{1 - b}}\left[\eta \tan^{-1}\sqrt{\frac{b}{1 - b}} + N\frac{\pi}{2}\right] \qquad (m = 1, 2, 3) \tag{7.23}$$

where:

$$\eta = \begin{cases} 1 & \text{(TE modes)} \\ (n_1/n_2)^2 & \text{(TM modes)} \end{cases}$$

Readers may derive similar relations for the TM modes by considering the boundary conditions for the electric field across the interface between the core and the cladding. In general, we would like to obtain b as a function of V, but Eq. (7.23) cannot be inverted to give an analytic expression. However, the functional relationship can be seen graphically by solving for V as a function of b and exchanging the ordinate and the abscissa to explicitly plot b as a function of V. Figure 7.3 shows the b–V curves found in this manner.

Let us discuss the TE_1 mode by referring to the curve in Fig. 7.3 labeled $N = 1$. When V reaches $\pi/2$ ($b \to 0$), the mode is cut off. In other words, when V is smaller than $\pi/2$, only the TE_0 mode can propagate. In terms of the waveguide parameters, this cutoff value is written as:

$$V_c = \frac{2\pi}{\lambda} a n_1 \sqrt{2\Delta} = \pi/2 \tag{7.24}$$

The single-mode condition for such a planar waveguide is $V < V_c$. For example, when $\lambda = 1.3\ \mu$m, $n_1 = 1.5$, and $\Delta = 0.01$, the maximum waveguide width $2a$ for single-mode transmission becomes 3.1 μm from the above condition. In the case of a semiconductor waveguide with $n_1 = 3.5$ and $\Delta = 0.05$, and single-mode width becomes 0.58 μm.

Figure 7.4 shows a curve of b versus V in the range of $V < \pi/2$, corresponding to the single-mode waveguide where only the TE_0 mode can propagate.

Figure 7.5 shows an electromagnetic field distribution of the TE_0 mode. Readers should notice the difference between this distribution and the func-

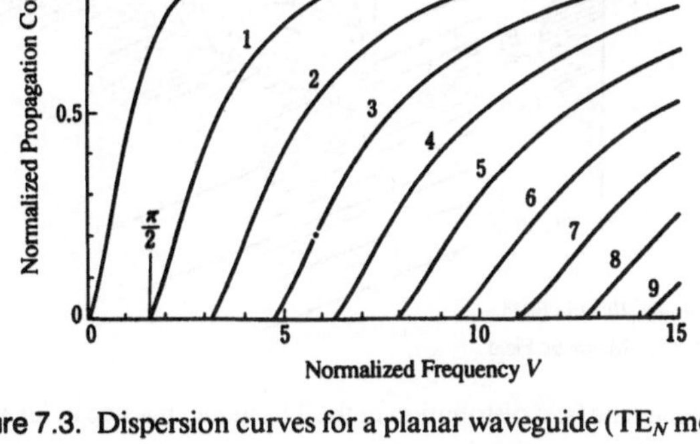

Figure 7.3. Dispersion curves for a planar waveguide (TE_N mode).

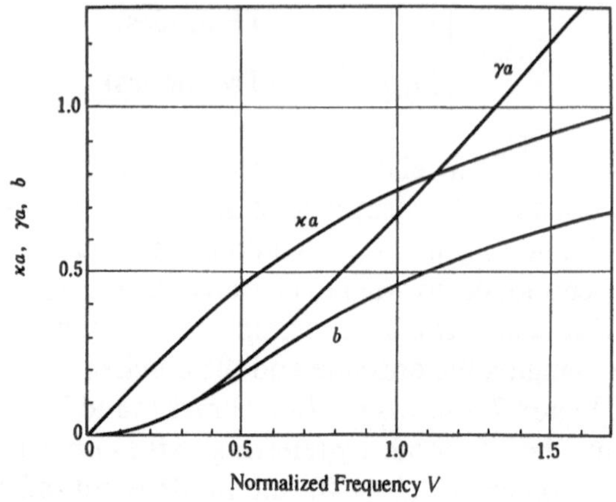

Figure 7.4. b–V curve of a single-mode waveguide where only the TE_0 mode can propagate.

tional relationships shown in Table 7.1. As seen from Eq. (7.23), b is characterized by the integral value N. This may give a hint regarding the question which was raised at the beginning of this chapter.

From among the mode functions in Table 7.1, the function in the core may be rewritten for even modes as:

$$\cos(\kappa x)\exp(-j\beta z) = (1/2)[\exp(j\kappa x - j\beta z) + \exp(-j\kappa x - j\beta z)] \qquad (7.24')$$

where each of the terms in the brackets denotes a plane wave. The propagation of a node in the plane wave is given by the expression, $j\kappa x - j\beta z = 0$; the propagation direction of the two plane wave terms in the brackets is written as:

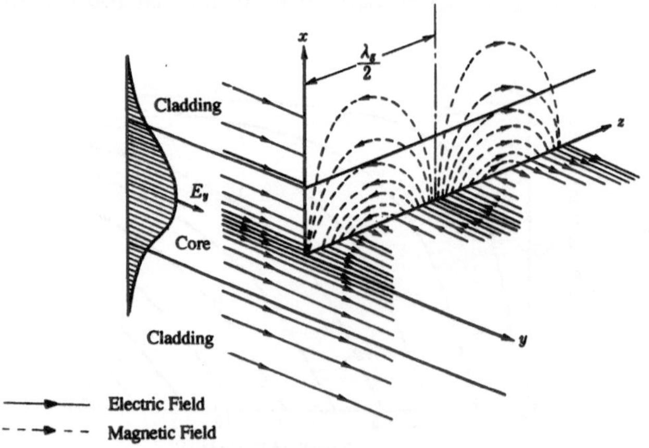

Figure 7.5. Electromagnetic field distribution of the TE_0 mode in a step index planar waveguide.[1]

$$\tan\theta = dz/dx = \pm\kappa/\beta \tag{7.25}$$

where θ is the angle to the optical axis. Since β and κ have discrete values, θ also has discrete value, θ_n, where n is an integer. κ can be written in terms of Eq. (7.23) as:

$$\kappa = (\pi/2a)[(2/\pi)\tan^{-1}(\sqrt{b/(1-b)}) + N] \tag{7.26}$$

From Eq. (7.22):

$$\beta = k_0 n_1[n_2/n_1 + b\Delta] \tag{7.27}$$

$$\therefore \tan\theta_N = (\pi/2k_0 n_1 a) \times [(2/\pi)\tan^{-1}(\sqrt{b/(1-b)}) + N]/[n_2/n_1 + b\Delta] \tag{7.28}$$

When $b \cong 1$ and Δ is much less than 1, the above expression reduces to:

$$\tan\theta_N \simeq (\pi/2k_0 n_2 a)(N+1) \tag{7.29}$$

As is clear from this expression, θ_N becomes large as N increases. It is interesting to note that even if $N = 0$, θ_N is not equal to zero. This is, all the light beams in a waveguide may be thought of as propagating with zigzag paths, and the higher modes (i.e., those with larger N's) have larger θ_N's.

7.2. MODES OF A THREE-DIMENSIONAL WAVEGUIDE

Let us now consider a three-dimensional waveguide, which has some structure or refractive index distribution in the lateral (horizontal) direction as well as in the vertical direction. An example is shown in Fig. 7.2b and 7.2c. To analyze this kind of waveguide, the so-called equivalent index method is sometimes very useful. This is based on the idea that we can use an equivalent index $n_{eq} = \beta/k_0$ [i.e., the local propagation constant β of the mode in the vertical direction (x direction) normalized by the wave number, k_0] as a new refractive index distribution. The ordinate in Fig. 7.3 shows the normalized propagation constant, b, and also shows the increment of the refractive index, n_2, normalized by $n_1 - n_2$. n_{eq} has values in the range of $n_2 \leq n_{eq} \leq n_1$. If the mode in the x direction is designated, the change in the waveguide thickness in the horizontal direction (y direction), and the index-of-refraction difference in that direction can be expressed in two dimensions as the refractive index distribution shown in Fig. 7.2d.

Let us first consider the case of Fig. 7.2b where the mode has a step index profile in the y direction, but the core thickness is not uniform. We can convert the problem into a two-dimensional waveguide with an index distribution. The mode distribution is derived as follows: In the case of a parabolic index of refraction and, further, where the width of the waveguide is relatively large, approximate analytic solutions can be utilized. Writing the distribution of the equivalent refractive index as:

$$n_{eq}^2(y) = \begin{cases} n_{eq}^2(0)[1 - (gy)^2] & (y \leqq W/2) \\ n_2^2 & (y > W/2) \end{cases} \tag{7.30}$$

$$g \simeq \frac{\sqrt{2(n_{eq}(0) - n_2)/n_{eq}(0)}}{W/2} \tag{7.31}$$

The mode function $u_p(y)$ becomes:

$$u_p(y) = \frac{1}{[2^p p! \sqrt{\pi} w_0]^{1/2}} H_p\left(\frac{y}{w_0}\right) e^{-(1/2)(y/w_0)^2} \tag{7.32}$$

where $H_p(x)$ is the Hermite polynomial. The characteristic spotsize, w_0, is evaluated as:

$$w_0 = \frac{1}{\sqrt{k_0 n_{eq}(0) g}} \tag{7.33}$$

The propagation constant β is approximated as follows:

$$\beta_p = \sqrt{k_0^2 n_{eq}^2(0) - k n_{eq}(0) g(2p + 1)} \simeq k_0 n_{eq}(0) - g(p + 1/2) \tag{7.34}$$

where, if $\beta_1 = k_0 n_{eq}$, the single-mode cutoff condition can be obtained.

If the normalized frequency in the y direction is given as follows:

$$V^e = k_0 n_{eq} \frac{W}{2} \sqrt{\frac{2(n_{eq} - n_2)}{n_{eq}}} \tag{7.35}$$

then the cutoff V value becomes $V_c^e = 3$. Equation (7.35) is an approximate expression and a more exact calculation gives $V_c^e = 2.4$.

For an arbitrary thickness and index-of-refraction distribution, a numerical calculation is required.

7.3. CONFINEMENT FACTOR

The mode confinement factor, ξ, is defined as the ratio of the fractional optical power confined in the core region to the total power. This factor is

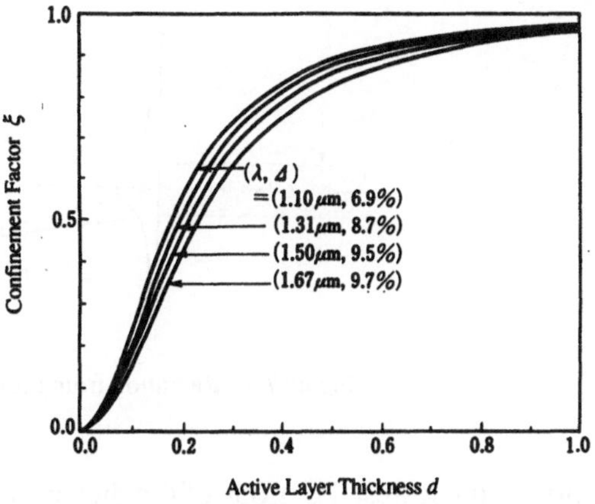

Figure 7.6. Mode confinement factor. λ, wavelength; Δ, normalized index difference. [After Y. Itaya, Y. Suematsu, S. Katayama, K. Kishino, and S. Arai, *Jpn. J. Appl. Phys.* **18**, 1795 (1979).][5]

important for a semiconductor laser having optical gain in the core region, because it is related to the mode gain. Using the above definition, the mode confinement factor of the even TE mode is calculated by an overlap integral of the active region and mode intensity distribution and found to be

$$\xi = \int_0^a |E(x)|^2 dx \Big/ \int |E(x)|^2 dx \tag{7.36}$$

Since the mode function was previously obtained in the case of the waveguide in Fig. 7.1a, the integral can be evaluated to give:

$$\xi = \frac{1 + 1/(2\kappa a)\sin(2\kappa a)}{1 + 1/(2\kappa a)\sin(2\kappa a) + 1/\gamma a \cos^2(\kappa a)} \tag{7.37}$$

With the help of some analytical calculations, the above expression can be reduced to a more general form[1]:

$$\xi = \frac{V + \sqrt{b}}{V + 1/\sqrt{b}} \tag{7.38}$$

The confinement factor ξ of a waveguide versus V, and the confinement factor ξ of a semiconductor laser versus the active layer thickness d (=2a), are shown in Figs. 7.4 and 7.6, respectively.

7.4. RADIATION FROM THE EDGE OF A WAVEGUIDE

Here we shall consider the radiation of light from the edge of a waveguide which is terminated vertically as shown in Fig. 7.7. The field distribu-

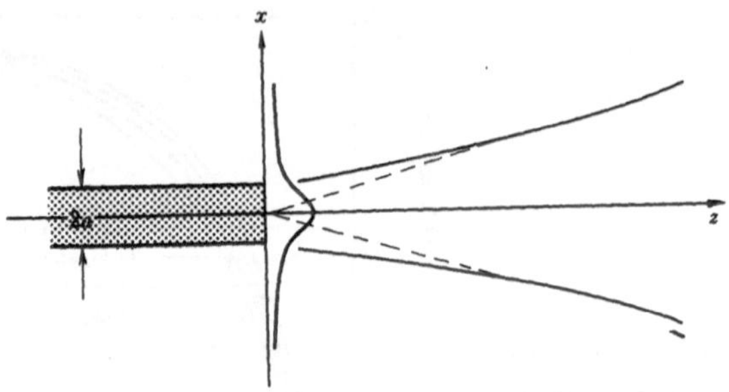

Figure 7.7. Radiation from the edge of a waveguide.

tion of the lowest TE mode of the dielectric waveguide is given from the two-dimensional approximation[3] as:

$$f_1(x, 0) = A_e \cos\kappa x \qquad (|x| \leqq a)$$
$$= A_e \cos\kappa a e^{-\gamma(|x|-a)} \qquad (|x| > a) \tag{7.39}$$

The radiation field distribution in free space is then obtained from the Fresnel–Kirchhoff integral developed in Chaper 6:

$$f_2(x, z) = \frac{j}{\lambda z} e^{-jkz} \int_{-\infty}^{\infty} dx' f_1(x', 0) \exp\left[-\frac{jk}{2z}(x - x')^2\right] \tag{7.40}$$

The rigorous solution can be obtained through numerical integration, but here we are going to generate an approximate form to give some physical insight. Let us consider the Gaussian beam equivalent to Eq. (7.39) which is written as:

$$f_g(x, 0) = A_e \exp\left[-\frac{1}{2}(x/s)^2\right] \tag{7.41}$$

(Fig. 7.8). If

$$|f_1(s_e, 0)|^2 = |f_g(s, 0)|^2$$

the equivalent spotsize, s_e, becomes:

$$s_e = a\left[1 + \frac{1}{\gamma a}\ln(\sqrt{e}\cos\kappa a)\right] \qquad (\cos\kappa a > 1/\sqrt{e})$$
$$= a\left[\frac{1}{\kappa a}\cos^{-1}(1/\sqrt{e})\right] \qquad (\cos\kappa a < 1/\sqrt{e}) \tag{7.42}$$

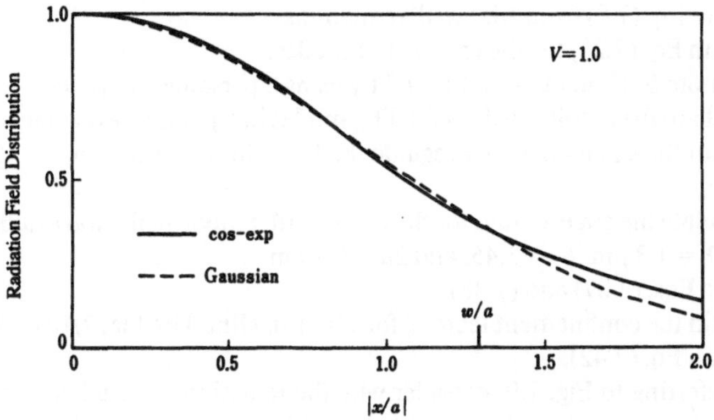

Figure 7.8. Equivalent Gaussian mode.

In most cases the width of the waveguide is thin and, therefore, Eq. (7.42) is meaningful. At a point far from the waveguide exit, the beam spreads by diffraction, and the spreading angle $2\Delta\theta$ is found to be:

$$2\Delta\theta = \frac{2}{ks} = \frac{2}{ka}\left[1 + \frac{1}{\gamma a}\ln(\sqrt{e}\cos\kappa a)\right]^{-1} \tag{7.43}$$

Figure 7.9 compares calculated and experimental data for the radiation angle from a semiconductor laser.

PROBLEMS

7.1. Obtain the critical angle θ_c for $\Delta = 0.5\%$ in degrees and radians.
7.2. Obtain the numerical aperture of a waveguide with $\Delta = 0.5\%$ and $n_1 = 3.5$.

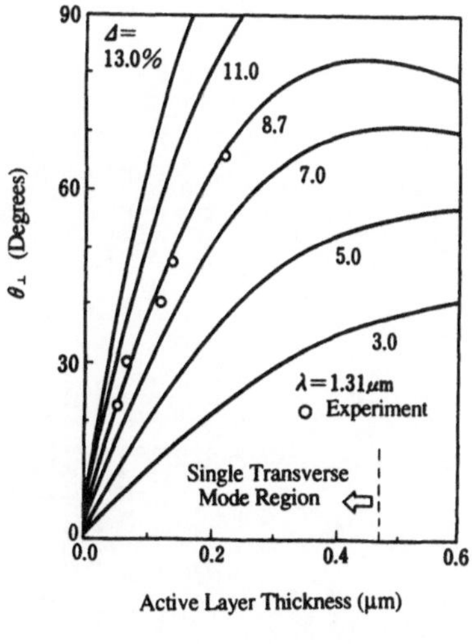

Figure 7.9. Radiation angle ($d = 2a$) from a semiconductor laser.[4,5] [After Y. Itaya, Y. Suematsu, S. Katayama, K. Kishino, and S. Arai, *Jpn. J. APpl. Phys.* **18,** 1795 (1979).][5]

7.3. Derive Eq. (7.9) from Maxwell's equations.
7.4. Obtain Eq. (7.23) for the case of TM modes.
7.5. Illustrate b–V curves as in Fig. 7.3 by using a personal computer.
7.6. What are the cutoff V values for TE_N modes in a planar waveguide?
7.7. Obtain the single-mode waveguide thickness for $\lambda = 1.55\ \mu m$, $\Delta = 5\%$, and $n_1 = 3.5$.
7.8. Calculate the angle of rays for the zeroth-order mode in the waveguide. Assume that $\lambda = 1.3\ \mu m$, $n_2 = 3.45$, and $2a = 0.2\ \mu m$.
7.9. Prove Eqs. (7.37) and (7.38).
7.10. Obtain the confinement factor ξ for $V = 1.0$. Hint: Use Fig. 7.4 and Eq. (7.38).
7.11. Obtain Eq. (7.42).
7.12. By referring to Fig. 7.9, consider why the radiation angle θ first increases and then decreases when the active layer thickness d increases. Hint: Remember the diffraction phenomenon.

REFERENCES

1. Y. Suematsu and K. Iga, *Introduction to Optical Fiber Communication,* revised 3rd ed., 3rd printing Ohmsha, Tokyo (1992).
2. K. Iga and Y. Kokubun, *Optical Fiber,* Ohmsha, Tokyo (1990).
3. H. Kogelnik, in *Integrated Optics* (T. Tamir, ed.), Springer-Verlag, Berlin (1979); *Guided-Wave Optoelectronics* (T. Tamir, ed.), Springer–Verlag, Berlin (1988).
4. K. Iga, *Appl. Opt.* **19,** 2940 (1980).
5. Y. Itaya, Y. Suematsu, S. Katayama, K. Kishino, and S. Arai, *Jpn. J. Appl. Phys.* **18,** 1795 (1979).
6. Y. Suematsu, ed., *Semiconductor Lasers and Integrated Optics,* Ohmsha, Tokyo (1984).
7. A. G. Fox and T. Li, *Bell Syst. Tech. J.* **40,** 453 (1961).
8. D. Marcuse, *Theory of Optical Waveguides,* Academic Press, New York (1974).
9. K. Iga, Y. Kokubun, and M. Oikawa, *Fundamentals of Microoptics,* Academic Press, New York (1984).

EIGHT

LASER RESONATORS AND RESONANT MODES

In this chapter, laser resonators and the concepts of normal modes and resonant frequencies are discussed, making use of the relationships obtained in Chapters 6 and 7. First, the resonant frequencies for a Fabry–Perot resonator consisting of a waveguide whose refractive index has a square law (parabolic) distribution will be presented. Second, the resonant modes and optical losses of a Fabry–Perot resonator which employs plain parallel or curved mirrors as reflectors are introduced. Finally, the physical principles and resonant frequencies of a distributed feedback resonator, which uses a diffraction grating, and a distributed Bragg reflector (DBR) resonator will be discussed.

8.1. INTRODUCTION

Laser oscillators and regenerative laser amplifiers require both amplification and feedback to operate. The amplification can be performed by a laser medium with the help of stimulated emission. An optical resonator is analogous to an electric resonant circuit in that it feeds back light to the amplifying medium and fulfills the well-known frequency condition for an oscillator. Since an electric circuit uses a discrete component, or a one-dimensional transmission line, the spatial distribution of the electromagnetic field is not important. On the other hand, at microwave and lightwave frequencies, three-dimensional resonators are utilized and we have to pay attention not only to the resonance frequency, but also to the spatial distribution of the electromagnetic field. This self-consistent field is called a "resonant mode" or "eigenmode." The purpose of this chapter is to introduce the general idea of a resonant mode, which is necessary for understanding the operation of a laser resonator. This will lead to a discussion and summary of the fundamental properties of a laser resonator.

The Fabry–Perot interferometer, which consists of a pair of parallel reflecting mirrors, is frequently utilized as an optical resonator.[1] In the first ruby laser, a pair of plain parallel mirrors was formed by coating the ends of the ruby rod.[2] A mathematical treatment of the resonant modes of a resonator with a pair of plain parallel mirrors of finite size was initially presented by Fox and Li.[3] A resonator with a pair of concave mirrors was later considered in order to reduce the diffraction losses. To represent the electric field mode in such a cavity, the so-called Gaussian mode was introduced.[4] Since that time, various types of Fabry–Perot optical resonators have been devised.

A semiconductor laser produces gain by stimulated emission from the recombination of electrons and holes which are injected into a *pn* junction. In the early development stages of semiconductor lasers, a homojunction was employed for the active region and some associated waveguiding was observed. To confine the light and the charge carriers more effectively in the active medium, the double heterostructure[6] laser was introduced in 1970, thus enabling continuous operation at room temperature. Usually, a Fabry–Perot resonator structure is obtained in semiconductor lasers by cleaving the crystals[7–9] or by etching.[10,11] Distributed feedback (DFB)[12] or distributed Bragg reflector (DBR)[13] are often employed in semiconductor lasers to achieve single-frequency operation. Since they do not require mirror facets, these structures are suitable for integrating optical waveguides and other photonic devices together with semiconductor lasers.

As mentioned previously, a laser oscillator requires both an active medium to amplify light and a resonator structure for feedback of the light. In semiconductor lasers the resonator consists of a waveguide to contain the optical mode in order to minimize optical loss, and a reflector for feedback. Table 8.1 shows a classification of some typical laser resonators.

Open-type resonators are used in gas and solid-state lasers and contain an active laser medium which is relatively independent of the mode formation, whereas the active medium of a semiconductor laser functions as a waveguide as well, and, therefore, forms an important component in the mode formation.

The dimensions of the resonators differ according to the type of laser. Usually, resonator dimensions for gas or solid-state lasers range from about 10 cm to several meters, and those for semiconductor lasers range from a few micrometers to 1 mm. In some extreme cases, such as in microcavity semiconductor lasers, the cavity length could be as small as $\lambda/2n$, where n is the refractive index of the material.

Table 8.1. Types of Laser Resonators

Model	Waveguide	Structure of reflector		Laser
Open resonator	Free space	Fabry–Perot type		Gas lasers
		Ring type		Solid-state lasers Dye lasers
Waveguide-type resonator	Hollow waveguide	Fabry–Perot type		Gas lasers
	Dielectric waveguide	Fabry–Perot type		Semiconductor lasers
		Distributed feedback type		
		Distributed Bragg reflector type		
Traveling wave type resonator	Active medium	One-side mirror		High-gain gas lasers

8.2. FABRY–PEROT WAVEGUIDE-TYPE RESONATORS

Let us first consider the resonance frequencies and modes in a distributed index-type waveguide laser resonator. In this case an analytic formulation is possible. Figure 8.1 shows a Fabry–Perot waveguide-type resonator whose length is L. The refractive index of the waveguide is assumed to be expressed by:

$$n^2(x, y) = n^2(0)[1 - g_1^2x^2 - g_2^2y^2] \tag{8.1}$$

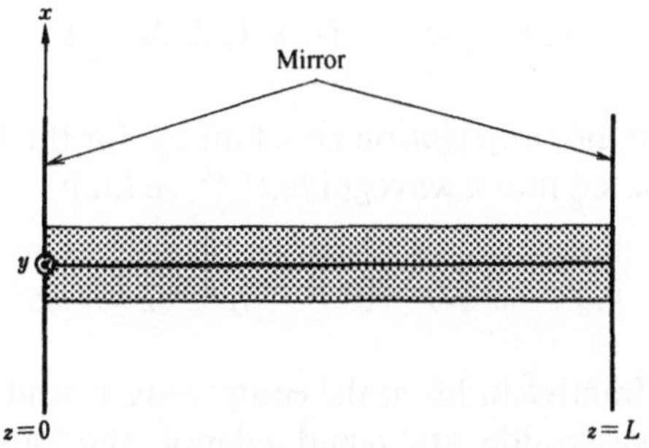

Figure 8.1. A model of a Fabry–Perot waveguide-type resonator.

In this resonator, the light is confined by the waveguiding effect of the index-of-refraction profile and is reflected at both end facets to make a standing wave as described below.

A guided wave propagating in the positive z direction inside the waveguide is expressed by:

$$\psi_{pg}^{+} = A_{pg}^{+} u_p(x, w_{01}) u_q(y, w_{02}) \exp[-j\beta_{pq} z] \tag{8.2}$$

A conjugate wave propagating in the negative z direction is:

$$\psi_{pq}^{-} = A_{pg}^{-} u_p(x, w_{01}) u_q(y, w_{02}) \exp[+j\beta_{pq} z] \tag{8.2'}$$

Therefore, the standing wave in the resonator can be written as:

$$\psi(x, y, z) = \sum_{p,q} (\psi_{pg}^{+} + \psi_{pq}^{-}) \tag{8.3}$$

$$= \sum_{p,q} \{A_{pq}^{+} \exp[-j\beta_{pq} z] + A_{pq}^{-}[+j\beta_{pq} z]\} u_p(x, w_{01}) u_q(y, w_{02}) \tag{8.4}$$

Assuming that the light is completely reflected at $z = 0$ and $z = L$, and the boundary condition is $\psi(x, y, 0) = \psi(x, y, L) = 0$, we can write:

$$A_{pq}^{+} + A_{pq}^{-} = 0 \tag{8.5}$$

Therefore,

$$\sin\beta_{pq} L = 0 \tag{8.6}$$

According to Eq. (8.6):

$$\beta_{pq} L = n\pi \qquad (n = 1, 2, 3, \ldots) \tag{8.7}$$

When we substitute the propagation constant β_{pq} for the Hermite–Gaussian mode of the distributed index waveguide,[14–18] we find:

$$\beta_{pq} = k_0 n(0) - (p + \tfrac{1}{2}) g_1 - (q + \tfrac{1}{2}) g_2 \tag{8.7'}$$

where $n(0)$ is the refractive index at the center axis, g_1 and g_2 are the focusing parameters of the waveguide, and p and q denote the transverse mode number in the x and y directions, respectively.

From Eq. (8.7), we can write, after some manipulation:

$$kL - (p + \tfrac{1}{2})g_1L - (q + \tfrac{1}{2})g_2L = n\pi \tag{8.8}$$

Since $k = n_1(\omega/c)$ and $\omega = 2\pi f$, the frequency, $f(n, p, q)$, which satisfies Eq. (8.8) is written as:

$$f(n, p, q) = \frac{c}{2n_1L}\left[n + \frac{g_1L}{\pi}\left(p + \frac{1}{2}\right) + \frac{g_2L}{\pi}\left(q + \frac{1}{2}\right)\right] \tag{8.9}$$

A corresponding equation for the wavelength, $\lambda(n, p, q) = c/f(n, p, q)$, may be written as:

$$\frac{\lambda(n, p, q)}{2n_1} = \frac{L}{\left[n + \frac{g_1L}{\pi}\left(p + \frac{1}{2}\right) + \frac{g_2L}{\pi}\left(q + \frac{1}{2}\right)\right]} \tag{8.10}$$

Since $g_1L, g_2L \ll 1$ in most cases, we can infer from Eq. (8.10) that the integer, n, denotes the number of half wavelengths $\lambda/2n_1(0)$ contained in the medium of length L. Therefore, the number n is called a longitudinal mode number. p and q, which denote the number of zeros of the mode function in the x and y directions, respectively, are called transverse mode numbers.

Next, let us find the frequency spacing Δf, or wavelength difference $\Delta\lambda$, for the case where the mode numbers, n, p, and q, are varied by some amount Δn, Δp, and Δq. From Eq. (8.9), we have:

$$\Delta f = \frac{\partial f}{\partial n}\Delta n + \frac{\partial f}{\partial p}\Delta p + \frac{\partial f}{\partial q}\Delta q \tag{8.11}$$

$$= \frac{c}{2n_1L}\Delta n + \frac{c}{2n_1}\cdot\frac{g_1}{\pi}\Delta p + \frac{c}{2n_1L}\cdot\frac{g_2}{\pi}\Delta q \tag{8.12}$$

The longitudinal mode spacing, Δf_1, can be obtained by putting $\Delta n = 1$:

$$\Delta f_1 = \frac{c}{2n_1L} \tag{8.13}$$

The transverse mode spacing, Δt, is found by setting Δp or $\Delta q = 1$:

$$\Delta f_t = \frac{cg_i}{2n_1\pi} \qquad (i = 1, 2) \tag{8.14}$$

For example, when L is 1 m and n_1 is 1, then Δf_1 is equal to 150 MHz.

The mode spacing in terms of the wavelength can be calculated in the same manner. The longitudinal mode separation, $\Delta\lambda_l$, is expressed as follows:

$$\Delta\lambda_l = \frac{\lambda^2}{2n_1 L} \tag{8.15}$$

The transverse mode spacing, $\Delta\lambda_t$, on the other hand, is given by:

$$\Delta\lambda_t = \frac{g_i\lambda^2}{2\pi n_i} \qquad (i = 1, 2) \tag{8.16}$$

As an example, suppose that $n_1 = 3.5$, $L = 300$ μm, $\lambda = 0.84$ μm, and $g_1 = 10^3$ (1/m), then $\Delta\lambda_l$ is 3.4 Å and $\Delta\lambda_t$ is 0.32 Å. Figure 8.2 shows a typical output spectrum of a semiconductor laser. The series of highest peaks represents a longitudinal mode group with a wavelength spacing of approximately 3.4 Å. The series of lower peaks seen just to the right of these represent another transverse mode which is approximately 0.3 Å from the fundamental transverse mode.

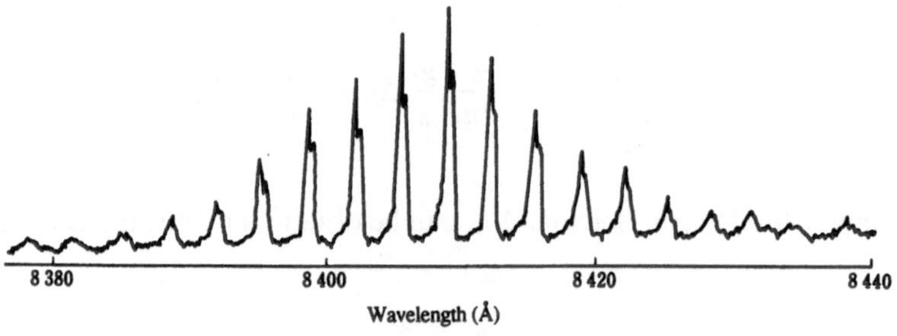

Figure 8.2. Spectrum of a semiconductor laser (GaAs laser).

8.3. OPEN FABRY–PEROT RESONATORS WITH CONCAVE MIRRORS

8.3.1. Spotsize

Figure 8.3 shows some configurations of Fabry–Perot resonators using various reflecting mirror shapes. Among these resonator configurations, the one with the plane and concave mirror (b) is the most widely used for gas or solid-state lasers. The so-called unstable resonator which utilizes a convex mirror, as shown in (d), has high diffraction loss since its resonant mode is leaky. This type of resonator is employed in relatively high gain lasers such as the solid-state Nd:YAG laser where the optical field spreads to fill the entire volume of the resonator, and more effective energy extraction is achieved. This kind of resonator was originally developed for the carbon dioxide laser and was later used for the ruby laser and YAG laser.

In Fig. 8.3b, a wave front is assumed to be a plane at $z = 0$. The wave-front coefficient, P_1, is expressed as:

$$P_1 = \frac{1}{s^2} \tag{8.17}$$

Here, s is the spotsize which is defined as the radius of the Gaussian beam where the power is $1/e^2$ of the value at the center. Using the F matrix which was introduced in Chapter 6, a total F matrix expressing the propagation

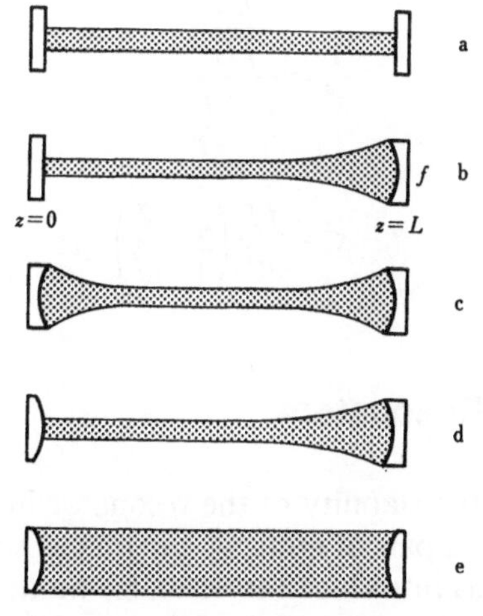

Figure 8.3. Fabry–Perot resonator (open resonator).

into free space from $z = 0$ to L, the reflection from the concave mirror, and the propagation in free space from $z = L$ to 0 can be expressed as:

$$\tilde{F} = \begin{pmatrix} 1 & 0 \\ j\frac{L}{k} & 1 \end{pmatrix} \begin{pmatrix} 1 & j\frac{k}{f} \\ 0 & 1 \end{pmatrix} \begin{pmatrix} 1 & 0 \\ j\frac{L}{k} & 1 \end{pmatrix}$$

$$= \begin{pmatrix} 1 - \frac{L}{f} & j\frac{k}{f} \\ j\frac{L}{k}\left(2 - \frac{L}{f}\right) & 1 - \frac{L}{f} \end{pmatrix} \equiv \begin{pmatrix} A & B \\ C & D \end{pmatrix} \tag{8.18}$$

According to the resonance condition, the wave-front coefficient after one round-trip must be the same as its initial value at $z = 0$. Thus, we can write:

$$P_1 = \frac{(AP_1 + B)}{(CP_1 + D)}$$

Then we have:

$$P_1 = \sqrt{\frac{B}{C}} \tag{8.19}$$

Accordingly,

$$\frac{1}{s^4} = \frac{\frac{k^2}{fL}}{2 - \frac{L}{f}}$$

$$s^4 = \frac{fL}{k^2}\left(2 - \frac{L}{f}\right) \tag{8.20}$$

8.3.2. Stability of Resonators

Let us consider the stability of the resonator in terms of how well we confine the light with a pair of concave mirrors as shown in Fig. 8.3c. The stability diagram[18] was introduced to visualize mode characteristics. Let us denote the radius of curvature of the left and right mirrors as b_1 and b_2,

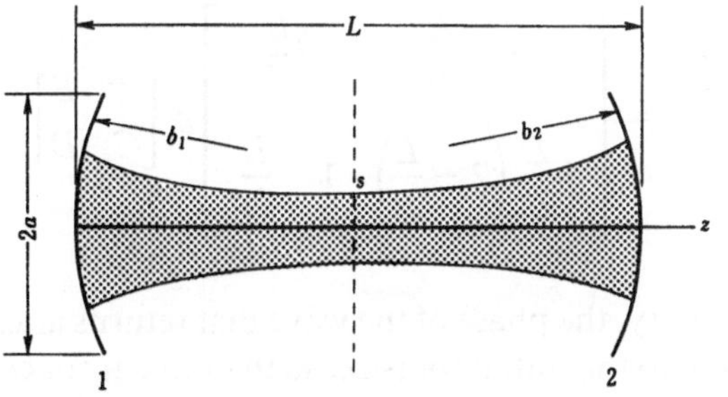

Figure 8.4. Fabry–Perot resonator with a pair of concave mirrors.

respectively, and the cavity length, L, as the distance between the mirrors, as shown in Fig. 8.4. The stability of the resonator can then be expressed in terms of a stability diagram as shown in Fig. 8.5. This diagram can be calculated from the matrix F mentioned above.

As an example, let $b_1 = b_2 = bf$ (f = focal length of the mirrors) and find the spotsize in the center of the resonator cavity and at one of the reflecting mirrors. The matrix F associated with the waves starting from the center of the resonator and returning again, after being reflected by mirror number 2, is expressed with the help of the matrices for free space propagation and reflection from a concave mirror:

$$\tilde{F} = \begin{bmatrix} 1 & 0 \\ j\dfrac{L}{2k} & 1 \end{bmatrix} \begin{bmatrix} 1 & j\dfrac{k}{f} \\ 0 & 1 \end{bmatrix} \begin{bmatrix} 1 & 0 \\ j\dfrac{L}{2k} & 1 \end{bmatrix}$$

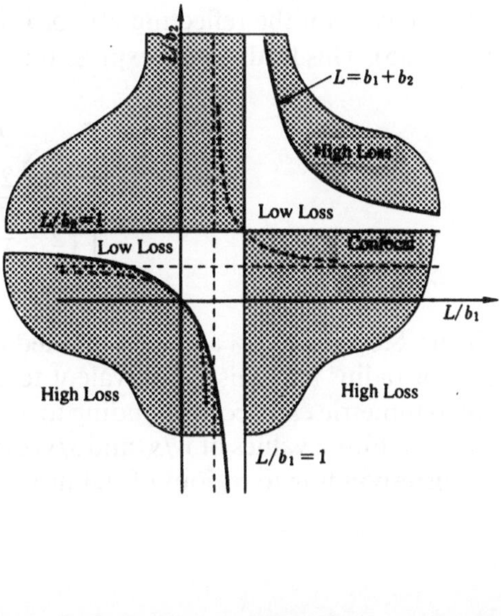

Figure 8.5. Stability diagram. [After G. D. Boyd and H. Kogelnik, *Bell Syst. Tech. J.* **41,** 1347 (1962).]

$$= \begin{bmatrix} 1 - \dfrac{L}{2f} & j\dfrac{k}{f} \\ j\dfrac{L}{2k}\left(2 - \dfrac{L}{2f}\right) & 1 - \dfrac{L}{2f} \end{bmatrix} = \begin{bmatrix} A & B \\ C & D \end{bmatrix} \tag{8.21}$$

Because of symmetry, the phase of the wave that returns after being reflected is plane and equal to the initial spotsize, so the wave-front coefficient, P, can be written as:

$$P = \frac{AP + B}{CP + D} \tag{8.22}$$

Note from Eq. (8.21) that $A = D$, so $P = \sqrt{B/C}$. Since $P = 1/s^2$, s^2 can be expressed as:

$$s^2 = \frac{1}{k}\sqrt{\frac{fL}{2}\left(2 - \frac{L}{2f}\right)} \tag{8.23}$$

When $L = 2f$, s achieves its maximum value, $s_0 = \sqrt{f/k}$, which is called the confocal condition. Normalizing Eq. (8.23) yields the following:

$$\frac{s}{s_0} = \left[\frac{L}{2f}\left(2 - \frac{L}{2f}\right)\right]^{1/4} \tag{8.24}$$

The spotsize on the reflecting mirror is obtained by substituting $z = L/2$ in Eq. (6.65). This leads to the expression:

$$\frac{w}{s_0} = \left[\frac{4 \times \dfrac{L}{2f}}{\left(2 - \dfrac{L}{2f}\right)}\right]^{1/4} \tag{8.25}$$

Figure 8.6 shows s/s_0 and w/s_0 plotted against $L/2f$. Since f is equal to the mirror radius $b/2$, this is equivalent to a plot of the spotsize versus L/b for the symmetric case, corresponding to a 45° line passing through the origin in Fig. 8.5. Finite values of w/s_0 and s/s_0 exist only in the region between 0 and 2, corresponding to regions of stability as shown in that diagram.

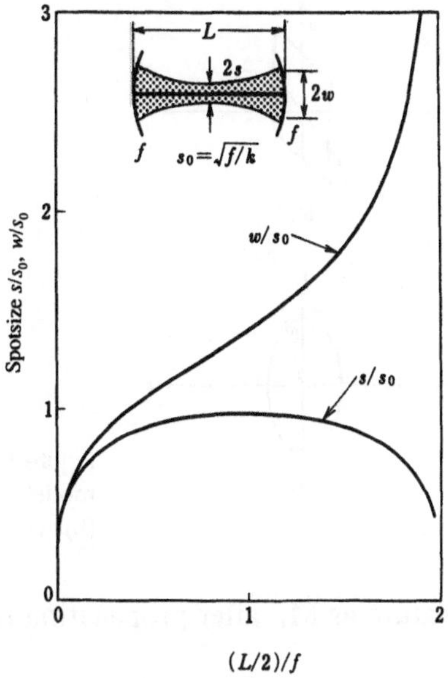

Figure 8.6. Spotsize in a Fabry–Perot resonator.

8.3.3. Mode and Diffraction Loss in Fabry–Perot Resonators

When the spotsize on a reflecting mirror is much smaller than the radius of that mirror, then the leakage of light around the reflecting mirror can be ignored and the mode field is similar to that found assuming an infinite mirror. On the other hand, when the size of the reflecting mirror becomes comparable to the spotsize, the perturbation of the resonant mode may become significantly different from the infinite mirror approximation.

To estimate the diffraction at a mirror, let us consider the mode which occurs in the case where the size of the reflecting mirror is finite, as shown in Fig. 8.7, where the two mirrors have the same curvature. The asymmetric case may be generalized by a not too difficult process.

8.3.3.1. Analysis by Cartesian Coordinates

The field distribution on the reflecting mirror M_1 is expressed as follows:

$$f_1(x, y) = \varphi(x)e^{+j(k/2R)x^2}\psi(y)e^{+j(k/2R)y^2} \tag{8.26}$$

where R is the radius of curvature of the mirror M_1.

Then the light that reaches the reflecting mirror M_2, which is assumed

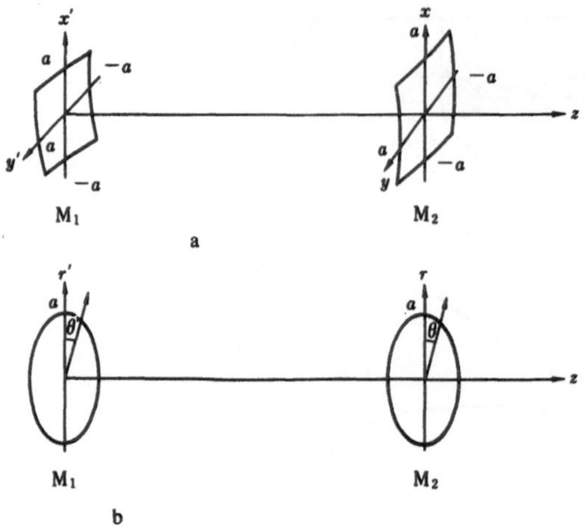

Figure 8.7. Fabry–Perot resonator models. (a) Cartesian coordinate; (b) cylindrical coordinate.

to have the same curvature as M_1, after propagating in the resonator is expressed as:

$$f_2(x, y) = \frac{j}{\lambda L} e^{-jkL} \iint f_1(x', y') e^{-(jk/2R)[(x-x')^2+(y-y')^2]} dx'dy' \tag{8.27}$$

Since $f_1(x', y')$ and $f_2(x, y)$ are resonant modes, we must write:

$$f_2(x, y) = \gamma\gamma' f_1^*(x, y) e^{-jkL} \tag{8.28}$$

where γ and γ' are complex numbers. This is because a self-consistent mode is formed in the cavity.

From the above equations we find:

$$\varphi(x)\psi(y) = \frac{1}{\gamma\gamma'} \cdot \frac{j}{\lambda L} \iint \varphi(x')\psi(y') e^{+j(k/L)(xx'+yy')} dx'dy' \tag{8.29}$$

By separating Eq. (8.29) into two independent equations with respect to x and y, the following integral equation for x can be obtained:

$$\varphi(x) = \gamma^{-1}\sqrt{j/\lambda L} \int \varphi(x') \exp\left[j\frac{k}{L} ss'\right] dx' \tag{8.30}$$

A similar equation can be obtained for y:

$$\sqrt{\frac{j}{\lambda L}} \cdot \sqrt{\frac{Lj}{k}} = \sqrt{j/2\pi} \tag{8.31}$$

Normalizing by $\sqrt{L/k}$ we can express these integral equations using the normalized variables $\xi = x/\sqrt{L/k}$ and $\xi' = x'/\sqrt{L/k}$:

$$\varphi(\xi) = C \int_{-\sqrt{2\pi N}}^{\sqrt{2\pi N}} \varphi(\xi') \exp[j\xi\xi']d\xi' \tag{8.32}$$

where

$$C = \gamma^{-1}\sqrt{\frac{j}{2\pi}} \tag{8.33}$$

and

$$N = a^2/\lambda L \tag{8.34}$$

The limits have been added to account for the finite size of the square mirror as shown in Fig. 8.7. In the case where $N \gg 1$, we may consider the lower and upper bounds to be $-\infty$ and $+\infty$, respectively. Then we can write:

$$\varphi = C \int_{-\infty}^{\infty} \varphi(\xi') \exp[j\xi\xi']d\xi' \tag{8.35}$$

If $\phi(\xi)$ is an even function, the following equation can be obtained:

$$\varphi^{(E)}(\xi) = 2C^{(E)} \int_0^{\infty} \varphi^{(E)}(\xi') \cos(\xi\xi')d\xi' \tag{8.36}$$

This equation can be solved by using an integral transform of Hermite polynomials:

$$\int_0^{\infty} \exp(-x^2) \cos 2\beta_x H_{2n}(\alpha x)dx = \frac{1}{2}\sqrt{\pi}(1-\alpha^2)^n \exp(-\beta^2) H_{2n}\left(\frac{\alpha\beta}{\sqrt{\alpha^2-1}}\right) \tag{8.37}$$

If we put $\alpha = \sqrt{2}$, $\sqrt{2}x = \xi$, and $\sqrt{2}\beta = \xi$, we can write

$$\int_0^\infty \exp\left(-\frac{1}{2}\xi'^2\right)H_{2n}(\xi)\cos(\xi\xi')d\xi' = \sqrt{\pi/2}\cdot j^{2n}\exp\left(-\frac{1}{2}\xi^2\right)H_{2n}(\xi) \quad (8.38)$$

Accordingly, the solution is

$$\varphi(\xi) = e^{(-1/2)\xi^2}H_{2n}(\xi) \quad (8.39)$$

$$\frac{1}{2C} = \frac{\sqrt{2}}{2}\sqrt{\pi}j^{2n} \quad (8.40)$$

$$\begin{aligned}\gamma &= \frac{1}{C}\sqrt{\frac{j}{2\pi}} = \sqrt{2\pi}(j)^{2n}\sqrt{\frac{j}{2\pi}} = j^{2n}\sqrt{j}\\ &= j^{2n}\sqrt{j}\\ &= e^{j(\pi/4+n\pi)}\\ &= e^{j\pi(n+1/4)}\end{aligned} \quad (8.41)$$

By combining terms, the following expression for the even function is obtained:

$$\varphi^{(E)}(\xi) = \frac{1}{[2^{2n}(2n)!\sqrt{\pi}]^{1/2}}e^{(-1/2)\xi^2}H_{2n}(\xi) \quad (8.42)$$

$$\gamma = e^{j\pi(n+1/4)} \quad (8.43)$$

where $\phi^{(E)}(\xi)$ is normalized.

Similarly, for the y component, we find:

$$\psi(\eta) = \frac{1}{[2^{2m}(2m)!\sqrt{\pi}]^{1/2}}e^{(-1/2)\eta^2}H_{2m}(\eta) \quad (8.44)$$

$$\eta = y/\sqrt{L/k} \quad (8.45)$$

$$\gamma' = e^{j\pi(m+1/4)} \quad (8.46)$$

$$\gamma^2 = \gamma\gamma' = e^{j\pi(n+m+1/2)} = e^{j(\pi/2)(2n+2m+1)} \quad (8.47)$$

The spotsize, s, is expressed as:

$$s = \sqrt{L/k} = \sqrt{f/k} \quad (8.48)$$

If $\psi(\xi)$ is an odd function, the following Hermite polynomial function may be used:

$$\int_0^\infty \exp\left(-\frac{1}{2}\xi^2\right)H_{2n-1}(\xi)\sin\xi\xi' d\xi' = \sqrt{\frac{\pi}{2}}(-1)^{2n}\exp\left(-\frac{1}{2}\xi^2\right)H_{2n-1}(\xi) \quad (8.49)$$

If we put

$$2n = p, \qquad 2m = q \quad (8.50)$$

then the eigenvalue of the integral equation is given by:

$$\gamma = e^{j(\pi/2)(p+q+1)} \quad (8.51)$$

where the new integral numbers p and q have been used to represent the mode number, whenever it is even or odd. The Hermite–Gaussian mode function describing the distribution in the resonator is shown below.

Hermite–Gaussian mode: HG_{nm}

$$\psi_{pq}(x, y, z) = N_{pq}\left(\frac{s}{w}\right)H_p\left(\frac{x}{w}\right)H_q\left(\frac{y}{w}\right) \times \exp\left[-\frac{1}{2}\left(\frac{1}{w^2}+j\frac{k}{R}\right)(x^2+y^2)\right] \times e^{-jkz+j(p+q+1)\varphi} \quad (8.52)$$

$$N_{pq} = \left(\frac{1}{2^p p! 2^q q! \pi}\right)^{1/2} \quad (8.53)$$

Here, $H_p(\ )$ is a Hermite polynomial of pth order and the spotsize, s, at the center of the resonator can be obtained from Eq. (8.24). The spotsize, s, at an arbitrary distance, the radius of curvature R of the wave front, and the

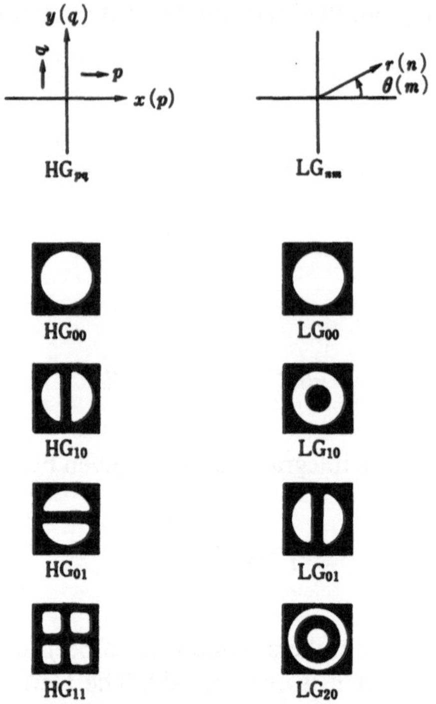

HG_{pq} : Hermite-Gaussian Mode
LG_{nm} : Laguerre-Gaussian Mode

Figure 8.8. Hermite–Gaussian modes.

phase shift, ϕ, can be calculated. In this case we choose $z = 0$ at the center of the resonator. Some mode patterns are illustrated in Fig. 8.8.

Next, let us consider the case where the size of the reflecting mirror cannot be ignored compared with a spotsize of the beam. In this case, an integral equation taking into account the lower and upper bounds of the integral must be solved. The eigenvalue γ can be obtained approximately using a variational method. That is, by multiplying both sides of Eq. (8.32) by $\phi(x)$, the following value of γ is obtained:

$$\gamma = \sqrt{\frac{j}{2\pi}} \cdot \frac{\int_{-\sqrt{2\pi N}}^{\sqrt{2\pi N}} \int_{-\sqrt{2\pi N}}^{\sqrt{2\pi N}} \varphi(\xi)\varphi(\xi') \exp[j\xi\xi'] d\xi d\xi'}{\int_{-\sqrt{2\pi N}}^{\sqrt{2\pi N}} \varphi^2(\xi) d\xi} \tag{8.54}$$

This integral is evaluated by substituting a suitable trial function for $\varphi(\xi)$. In general, a good approximation is obtained by using the solution for $N \rightarrow \infty$ which gives an analytic expression.

8.3.3.2. Analysis by Cylindrical Coordinates

To obtain the eigenmode by cylindrical coordinates, as shown in Fig. 8.4, the diameter of the mirrors is assumed to be $2a$. The electric field distribution on the mirror, M_1, is taken as:

$$\psi_1(r, \theta) = S_m(r)\mathrm{e}^{+j(k/f)r^2}\mathrm{e}^{-jm\theta} \tag{8.55}$$

The light that is reflected by the mirror, M_2, after propagation across the resonator is expressed as:

$$\psi_2(r, \theta) = \frac{j}{\lambda L}\mathrm{e}^{-jkL}\iint r'dr'd\theta'\psi_1(r', \theta') \times \exp\left[-\frac{jk}{2L}\{r'^2 - 2rr'\cos(\theta' - \theta) + r^2\}\right] \tag{8.56}$$

For ψ_1 and ψ_2 to be resonant modes, ψ_2 must match ψ_1^*, except for some complex constant and phase change assuming a symmetric cavity. Therefore, the following condition must be satisfied:

$$\psi_2(r, \theta) = \gamma^2\psi_1^*(r, \theta)\mathrm{e}^{-jkL} \tag{8.57}$$

Then,

$$S_m(r)\mathrm{e}^{+jm\theta} = \gamma^{-2}\cdot\frac{j}{\lambda L}\iint r'dr'd\theta' S_m(r')\mathrm{e}^{-jm\theta'} \times \exp\left[+j\frac{k}{L}rr'\cos(\theta - \theta')\right] \tag{8.58}$$

where we have assumed $2f = L$ to simplify the equation. This confocal condition corresponds to the smallest diffraction loss. When we integrate with respect to ϕ', we obtain the following integral equation which contains a symmetric kernel[3]:

$$S_m(r)\sqrt{r} = \gamma_m^{-2} \int_0^a \sqrt{r'}dr' K_m(r, r') S_m(r')$$

$$K(r, r') = j^{n+1} \frac{k}{L} J_m\left(k \frac{rr'}{d}\right) \sqrt{rr'} \quad (8.59)$$

where a is the mirror radius, $s_{nm}(r)$ is the eigenfunction and γ_{nm}^2 is the eigenvalue of the integral equation, and n (=0, 1, 2, . . .) denotes the mode order in the radial direction. From the definition of the constant γ, the diffraction loss α_{nm} and the phase shift ϕ_{nm} are given as:

$$\alpha_{nm} = 1 - |\gamma_{nm}^2|^2 \quad (8.60)$$

$$\varphi_{nm} = \arg(\gamma_{nm}^2) \quad (8.61)$$

Fox and Li[3] found the eigenfunctions and eigenmodes by solving the integral equation using an iterative integration. The mode in the Fabry–Perot resonator is approximately expressed by a transverse electromagnetic (TEM) wave. We represent the mode by TEM_{nm} where n and m correspond to the mode numbers. The diffraction loss and phase shift obtained by Fox and Li are shown in Figs. 8.9 and 8.10, respectively. The abscissa is the Fresnel number which is defined as:

$$N = \frac{a^2}{L\lambda} \quad (8.62)$$

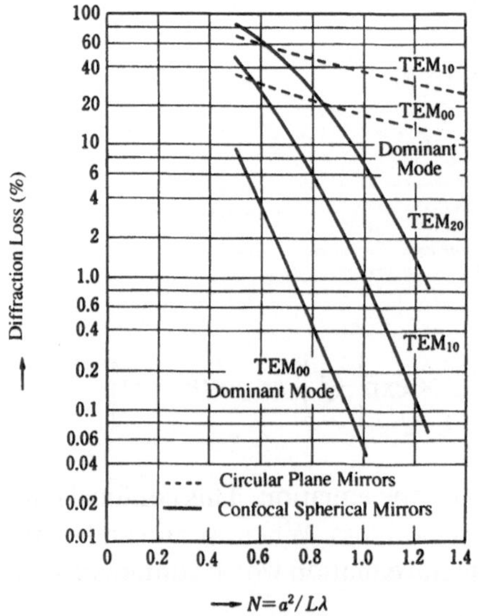

Figure 8.9. Diffraction loss/propagation. [After A. G. Fox and T. Li, *Bell Syst. Tech. J.* **40**, 453 (1961).]

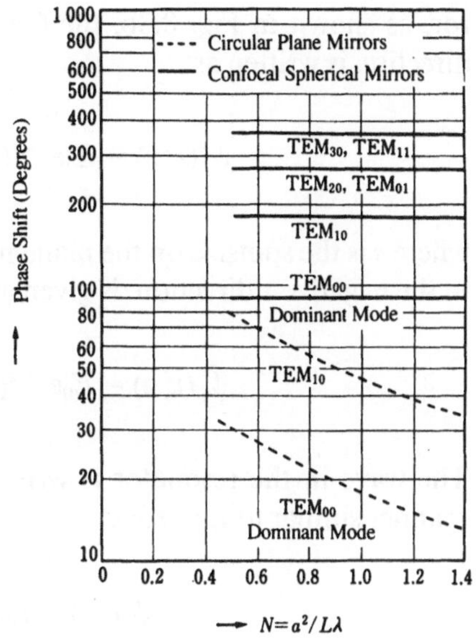

Figure. 8.10. Phase shift in the Fabry–Perot resonator. [After A. G. Fox and T. Li, *Bell Syst. Tech. J.* **40,** 453 (1961).]

This Fresnel number represents a normalized resonator in terms of wavelength, diameter of mirrors, and cavity length.

When the radius of the reflecting mirror, a, is considerably larger than the spotsize of the resonant mode, the integral equation [Eq. (8.59)] can be analytically solved by assuming $a \to \infty$. The solution is expressed using Laguerre–Gaussian functions in cylindrical coordinates.[4]

Laguerre–Gaussian mode: LG$_{nm}$

$$\psi_{nm}(r, \theta, z) = N_{nm}\left(\frac{s}{w}\right)\left(\frac{r}{w}\right)^m L_n^m\left(\frac{r^2}{w^2}\right)$$

$$\times\begin{pmatrix}\cos m\theta\\ \sin m\theta\end{pmatrix}\exp\left[-\frac{1}{2}\left(\frac{1}{w^2}+j\frac{k}{R}\right)r^2\right]\times e^{-jkz+j(2n-m+1)\varphi} \qquad (8.63)$$

$$N_{nm} = \left[\frac{(n-m)!}{(n!)^3\pi}\right]^{1/2} \qquad (8.63')$$

where $L_n^m(\ )$ is the associated Laguerre polynomial.

8.3.4. Resonance Frequency

We now seek to determine the resonance frequency in a Fabry–Perot resonator. When the resonator consists of a plane mirror and a concave mirror, as shown in Fig. 8.3b, the Gaussian beam traveling in the positive z direction is written as:

$$\psi^{+}(r, z) = E_0 e^{-jkz}(s/w)\exp\{-\tfrac{1}{2}P_r^2 + j\varphi] \tag{8.64}$$

where s is the spotsize on the plane mirror. The conjugate wave ψ^{-}, traveling in the negative z direction, is given as:

$$\psi^{-}(r, z) = E_0 e^{+jkz}(s/w)\exp[-\tfrac{1}{2}P_r^2 - j\varphi] \tag{8.65}$$

The wave in the resonator is written as the sum of these two waves in a manner similar to Eq. (8.4):

$$\psi(r, z) = E_0[A^{+}\psi^{+} + A^{-}\psi^{-}] \tag{8.66}$$

When $z = 0$, from the boundary condition $\psi = 0$, A^{-} becomes $-A^{+}$. Also, when $z = L$, from the boundary condition $\psi = 0$:

$$\sin[kL - \varphi] = 0 \tag{8.67}$$

From Eq. (6.68) for a propagating Gaussian beam, this gives:

$$kL - \tan^{-1}\left(\frac{L}{ks^2}\right) = n\pi \qquad (n = 1, 2, 3, \ldots) \tag{8.68}$$

Since s is determined from the spatial resonance condition, the following equation is obtained:

$$f_n = \frac{c}{2L}\left[n + \frac{1}{\pi}\tan^{-1}\sqrt{\frac{L/2f}{2 - L/f}}\right] \tag{8.69}$$

where the number n is a mode number of the longitudinal mode similar to the preceding section. Equation (8.69) corresponds to the case $p = 0$ and $q = 0$ in the transverse mode. When we extend the argument to the Hermite–Gaussian modes of pth or qth order, we can obtain an equation similar to Eq. (8.9):

$$f(n, p, q) = \frac{c}{2L}\left[n + \left(p + \frac{1}{2}\right)\frac{1}{\pi}\tan^{-1}\sqrt{\frac{L/f}{2 - L/f}}\right.$$
$$\left. + \left(q + \frac{1}{2}\right)\frac{1}{\pi}\tan^{-1}\sqrt{\frac{L/f}{2 - L/f}}\right] \quad (8.70)$$

From the above equation, the longitudinal mode spacing, Δf, is readily found to be:

$$\Delta f = c/2L \quad (8.71)$$

When $L = 2f$, $fm^{-1}\sqrt{(L/f)/(2 - L/f)}$ becomes $\pi/4$, and, therefore, the expression in the brackets on the right-hand side of Eq. (8.70) becomes $n + (p + q + 1)/4$. This causes a degeneracy for different combinations of pth and qth order modes. The resonator with such a confocal structure ($L = 2f$) becomes extremely unstable and may experience mode hopping between higher-order transverse modes. This sort of instability can be eliminated by offsetting the cavity length, L, to be a little bit different from $2f$.

8.4. DISTRIBUTED FEEDBACK/REFLECTOR RESONATORS

8.4.1. Resonance Frequencies

The distributed feedback (DFB) resonator is one in which a grating with a periodicity, Λ, is imposed on or near a waveguide as shown in Fig. 8.11.

Kogelnik and Shank[20] succeeded in achieving laser oscillation using this type of resonator in waveguides made from organic materials which were optically pumped. After that, Nakamura, Yariv, and co-workers[12] built a DFB resonator using semiconductor materials. This type of resonator produces a strong reflection when the scattered waves are added in-phase. This phenomenon occurs when the periodicity Λ is an integral multiple of $\lambda_g/2$, where λ_g is the wavelength of the light in the guide.

This type of resonator can be fabricated by monolithic processes.

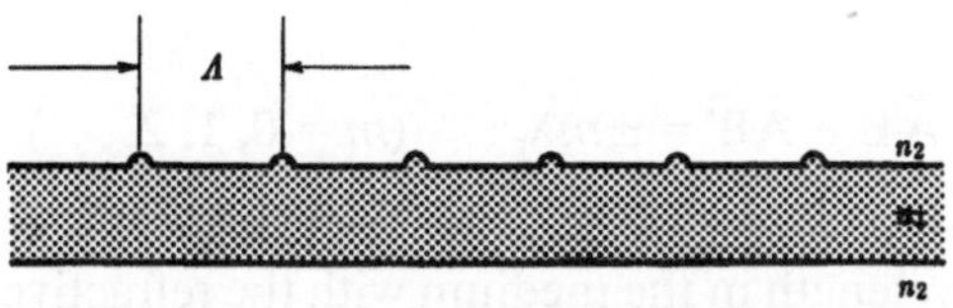

Figure 8.11. Distributed feedback (DFB) resonator.

Achieving single-frequency operation is an important aspect of a DFB resonator since the resonance wavelength depends on the periodicity Λ. For the reasons stated above, the resonance wavelength, λ_0, may be written:

$$\lambda_0 \cong 2\lambda n_{eq}/N \qquad (N = 1, 2, 3, \ldots) \tag{8.72}$$

where n_{eq} is the equivalent refractive index in the waveguide and λ_B is called the Bragg wavelength.

By differentiating Eq. (8.72), the mode spacing, $\Delta\lambda$, is written:

$$\begin{aligned}\Delta\lambda &= \frac{\partial\lambda_0}{\partial N}\Delta N + \frac{\partial\lambda_0}{\partial\lambda}\Delta\lambda \\ &= -\frac{\lambda_0}{N}\Delta N + \frac{\lambda_0}{n_{eq}}\frac{\partial n_{eq}}{\partial\lambda}\Delta\lambda \\ |\Delta\lambda| &= \frac{\lambda_0/N}{\left(1 - \dfrac{\lambda/_0}{n_{eq}}\dfrac{\partial n_{eq}}{\partial\lambda}\right)}\end{aligned} \tag{8.73}$$

8.4.2. Diffracted Waves

The diffraction grating acts as a source to diffract the transmitting lightwave. If the origin of the diffraction is periodic, the diffracted waves add together in a specific direction and strong scattering or diffraction occurs. Figure 8.12 shows the diffraction of light which is incident from the left bottom at an angle, θ_i, upon a waveguide-type diffraction grating. The light is scattered in various directions from the convex diffraction element lines. The diffraction angle is expressed as θ_m, and, from Fig. 8.12, the phase of the wave front is constant between A and A′ before scattering, and B and B′ after scattering. Thus, the distances A′B and AB′ must differ by an integer multiple of the wavelength:

$$\overline{A'B} - \overline{AB'} = \pm m\lambda_g \qquad (m = 0, 1, 2, \ldots) \tag{8.74}$$

Since λ_g is the wavelength in the medium with the refractive index n_2, we can write:

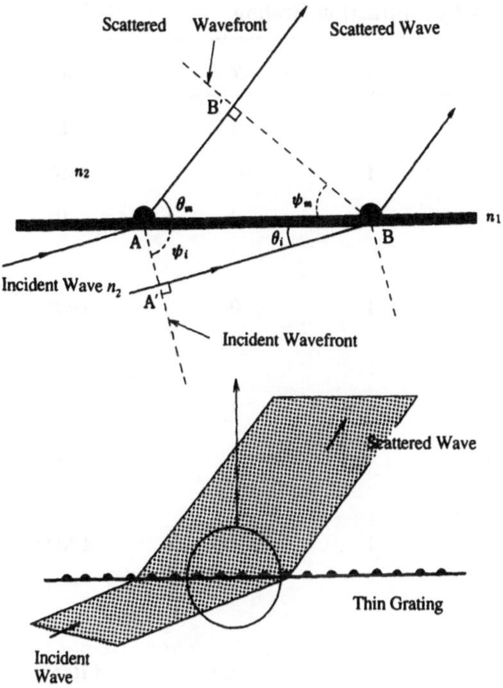

Figure 8.12. Scattering of waves from a waveguide grating.

$$\lambda_g = \lambda_0/n_2 \tag{8.75}$$

where λ_0 is the wavelength in a vacuum. We can also write:

$$\overline{A'B} = \Lambda \cos\theta_i \tag{8.76}$$

$$\overline{AB'} = \Lambda \cos\theta_m \tag{8.77}$$

From Eq. (8.72):

$$\Lambda = \lambda_0/(2n_{eq})\cdot N \tag{8.78}$$

Table 8.2 shows some examples of Λ's against N. $\lambda_0/(2n_{eq})$ has been chosen

Table 8.2. Relationship between Lattice Order N and Periodicity Λ

N	Λ (μm)
1	0.23
2	0.46
3	0.69

Table 8.3. Scattering by Diffraction Grating

N	$\cos\theta_m$	m	$\cos\theta$	θ	Direction
1	$1+2m$	0	1	0°	
		1	−1	180°	
2	$1+m$	0	1	0°	
		1	0	±90°	
		2	−1	180°	
3	$1+2m/3$	0	1	0°	
		1	1/3	±70.5°	
		2	−1/3	±109.5°	
		3	−1	180°	

to be 1.6/(2 × 3.5). Accordingly, Eq. (8.74) becomes:

$$\cos\theta_m = \cos\theta_i \pm \frac{2m}{N}\left(\frac{n_{eq}}{n_2}\right) \tag{8.79}$$

When $\theta_i = 0$ and $n_{eq} = n_2$:

$$\cos\theta_m = 1 \pm \frac{2m}{N} \tag{8.80}$$

Table 8.3 shows the grating order N and the diffraction angle to the diffraction order m. We must pay attention to the fact that light is scattered both along the waveguide and at other directions with angle θ_m.

8.4.3. Stop Bands

It was shown in Eq. (8.72) that the resonance frequency of the DFB resonator depends on the periodicity of the structure. It is known, however,

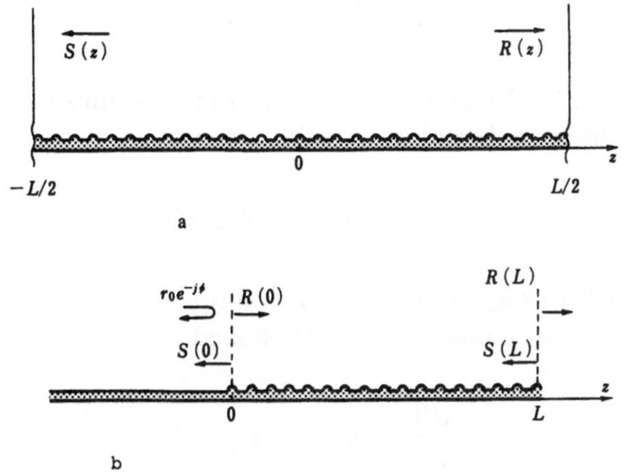

Figure 8.13. (a) Distributed feedback (DFB) resonator; (b) distributed Bragg reflector (DBR)-type resonator.

from a more precise calculation that the resonance diffraction does not occur at the Bragg wavelength λ_B. Kogelnik and Shank[20] predicted the existence of a stop band, at the center of which is located the Bragg wavelength. Figure 8.13 shows a waveguide with a diffraction grating. For simplicity, we assume that the equivalent refractive index n of the waveguide changes periodically as given by:

$$n = n_0 + \Delta n \cos(2\pi z/\Lambda) \tag{8.81}$$

where Λ is the pitch of the grating and Δn is the change of refractive index. It can easily be understood from the discussions in Chapter 7 that variations in the waveguide thickness may be expressed in terms of the change of the equivalent refractive index. So this equation is rather generally applicable. The couple coefficient κ is defined as:

$$\kappa = \pi \Delta n/\lambda_0 = k_0 \Delta n/2 \tag{8.82}$$

We also assume that the waveguide has a net gain (or loss). The electric field $E(z)$ in the waveguide satisfies the following wave equation:

$$\frac{\partial^2 E}{\partial z^2} + k^2 E = 0 \tag{8.83}$$

where

$$k^2 = k_0^2 n^2 + jk_0 ng \tag{8.84}$$

Now $\beta_0 = \pi/\Lambda$ is the Bragg frequency, and $E(z)$ is written as the sum of a forward wave and a backward wave as follows:

$$E(z) = R(z)e^{-j\beta_0 z} + S(z)e^{j\beta_0 z} \tag{8.85}$$

By substituting Eq. (8.85) into (8.83) and multiplying by either $e^{j\beta_0 z}$ or $e^{-j\beta_0 z}$, the following coupled equations can be obtained:

$$-\frac{dR}{dz} + (g/2 - j\delta)R = j\kappa S \tag{8.86}$$

$$\frac{dS}{dz} + (g/2 - j\delta)S = j\kappa R \tag{8.86'}$$

where

$$\delta = \beta - \beta_0 = k_0 n_0 - \pi/\Lambda \tag{8.87}$$

δ is the offset of the frequency from the Bragg frequency, and d^2R/dz^2 and d^2S/dz^2 are ignored. The solutions of the coupled equation (8.86) are written as:

$$R = \gamma_1 e^{\gamma z} + \gamma_2 e^{-\gamma z} \tag{8.88}$$

$$S = s_1 e^{\gamma z} + s_2 e^{-\gamma z} \tag{8.88'}$$

By substituting Eq. (8.88) into (8.86), the following relation is obtained:

$$\gamma^2 = \kappa^2 + (g - j\delta)^2 \tag{8.89}$$

Assuming that reflections at both ends of the boundary of the DFB can be ignored, we can write:

$$R(-L/2) = 0 \tag{8.90}$$

$$S(L/2) = 0 \tag{8.90'}$$

Then

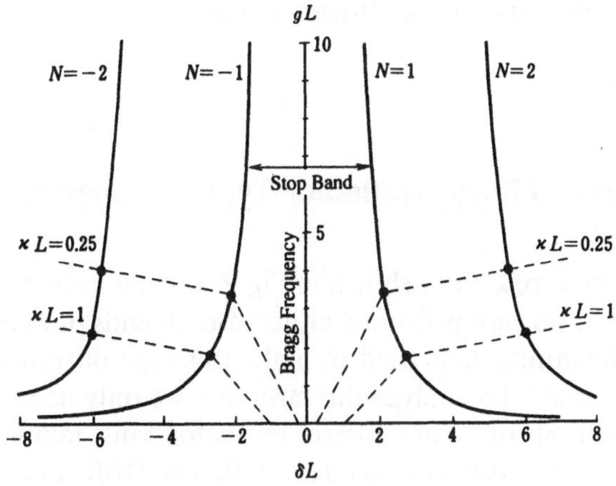

Figure 8.14. Eigenvalue of gL against δL.

$$r_1 = s_2, \qquad r_2 = s_1 \tag{8.91}$$

and

$$r_1/r_2 = s_2/s_1 = -e^{\gamma L} \tag{8.92}$$

Therefore, R and S are written as:

$$R(z) = \sinh\gamma(z - L/2) \tag{8.93}$$

$$S(z) = \sinh\gamma(z + L/2) \tag{8.93'}$$

By substituting Eqs. (8.93) and (8.93′) into Eqs. (8.86) and (8.86′), and rearranging the terms, the following equations are obtained:

$$\gamma + (g/2 - j\delta) = j\kappa e^{-\gamma L} \tag{8.94}$$

$$\gamma - (g/2 - j\delta) = -j\kappa e^{+\gamma L} \tag{8.94'}$$

From the sum and the difference of the above equations, we can write:

$$\kappa = j\gamma/\sinh\gamma L \tag{8.95}$$

$$g/2 - j\delta = \kappa \cosh\gamma L \tag{8.96}$$

The values of g and δ can be obtained by finding the value of γ which satisfies the condition and by substituting that value of γ into Eq. (8.96). Here κ in the eigenvalue equations is assumed to be a real number.

Figure 8.14 shows the eigenvalue gL plotted against δL. As seen from this figure the vicinity of the Bragg frequency, i.e., $\delta = 0$, becomes a stop band.

8.4.4. Distributed Bragg Reflector-Type Resonators

The DBR-type resonator shown in Fig. 8.13b has a structure in which a DBR region with no gain is formed either at both ends or at one end of the active layer. (Kaminow first tried to make this type of resonator by using an organic material.) To analyze this structure we only need to discuss the resonance conditions of a Fabry–Perot resonator while taking into account the reflectivity of the DBR and its phase shift. The DBR resonator allows us to independently design the active part of the laser and the DBR part, as well as the connection into an output waveguide at the end of the DBR. In the analysis of the DBR structure, the reflection at the end was ignored. If that reflection is considered, however, a more complicated boundary condition must be imposed.

To model the resonator we assume that the reflectivity on the right side, extending from $z = 0$ to $z = L/2$ in Fig. 8.13b, is expressed as follows:

$$r = r_0 e^{-j\phi} \tag{8.97}$$

We seek to obtain expressions for r_0 and ϕ. As before, we assume that the reflection at the right far end is ignored. That is, $s(L) = 0$.

The wave equation for the DBR is identical to that for the DFB discussed previously. We can use the results of that discussion to calculate the reflectivity and the phase shift as a function of the DFB parameters and the frequency offset, δ.

The complex reflectivity, r, is obtained by using Eq. (8.93) and the boundary conditions:

$$r = \frac{S(0)}{R(0)} = \frac{-j\kappa \tanh\gamma L}{\gamma + j\delta \tanh\gamma L} \tag{8.98}$$

where the gain $g = 0$ and γ is given by:

$$\gamma^2 = \kappa^2 - \delta^2 \tag{8.99}$$

In the region where $\delta < k$, γ is a real number and r_0 and ϕ become:

$$r_0 = \frac{\kappa \tanh\sqrt{\kappa^2 - \delta^2}L}{\sqrt{\kappa^2 - \delta^2} + \delta^2 \tanh\sqrt{\kappa^2 - \delta^2}L} \tag{8.100}$$

$$\phi = \frac{\pi}{2} + \tan^{-1}\left[\frac{\delta}{\sqrt{\kappa^2 - \delta^2}} \tanh\sqrt{\kappa^2 - \delta^2}L\right] \tag{8.101}$$

At the Bragg frequency ($\delta = 0$), r_0 and ϕ simplify to:

$$r_0(0) = \tanh\kappa L \tag{8.102}$$

$$\phi(0) = \frac{\pi}{2} \tag{8.103}$$

For example, if $\kappa L = 2$, $|r_0|^2 = 0.92$.

8.4.5. $\lambda_B/4$ Phase Shift

As is seen in the preceding section, the phase shift of the DBR at the Bragg frequency is $\pi/2$. If we add another DBR symmetrically to the left side, the device becomes a DFB resonator. The phase shift of the left DBR is also $\pi/2$, so the total round-trip phase shift becomes π, leading to a stop band. When a region with a phase shift, ψ, is inserted as illustrated in Fig. 8.15, the total phase shift for one round-trip through this section becomes 2ψ. Consequently, the total phase shift required for oscillation at the Bragg frequency is written as:

$$\underset{(\text{DBR}①)}{2\psi + \pi/2} + \underset{(\text{DBR}②)}{\pi/2} = 2\pi q \qquad (q = 1, 2, 3, \ldots) \tag{8.104}$$

The minimum phase shift corresponds to $q = 1$, or $\phi = \pi/2$.

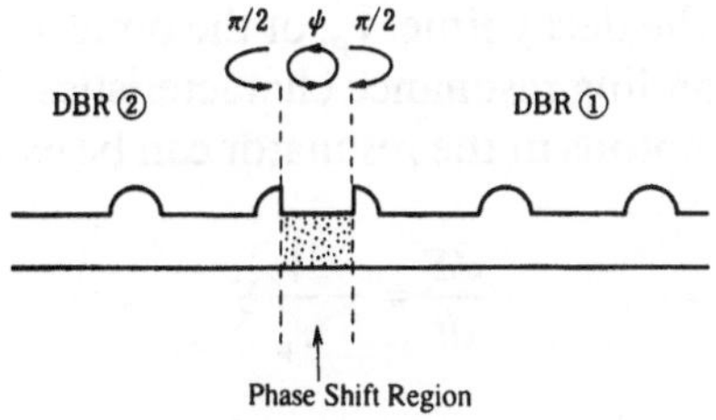

Figure 8.15. Phase-shifted DFB.

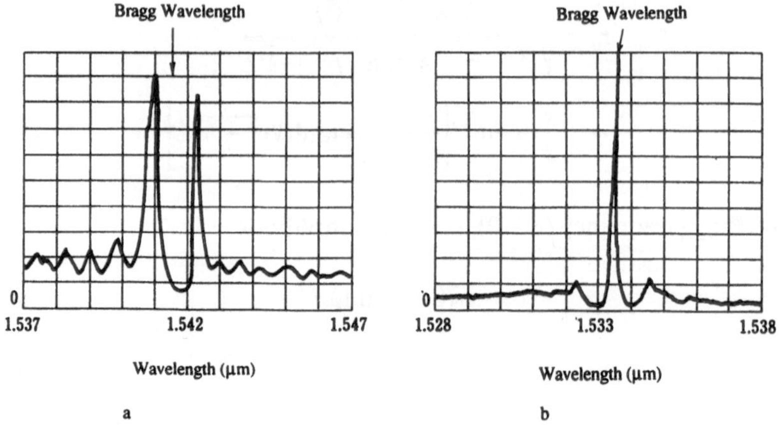

Figure 8.16. Spectra of a DFB laser. (a) Homogeneous DFB; (b) $\lambda/4$-shifted DFB. [After K. Utaka, S. Akiba, K. Sakai, and Y. Matsushita, *IEEE J. Quantum Electron.* **QE-22,** (July 1986).]

The length of the inserted section giving this phase shift is equivalent to a quarter wavelength. For this reason the phase-shifting region sandwiched in the middle of two DBRs is called a $\lambda_B/4$ shifter. Equation (8.104) shows that the phase-shifting region may be odd multiples of $\lambda_B/4$ or it may be a slightly different length with a slightly different propagation constant.

Figure 8.16b shows a subthreshold spectrum of a DFB laser with a $\lambda/4$ phase shifter. We can see that the resonance occurs at the Bragg frequency λ_B. In contrast to Fig. 8.16b, Fig. 8.16a shows the spectrum of a conventional homogeneous DFB laser which exhibits a stop band and two wavelengths are resonant on either side of the Bragg wavelength.

8.5. RESONATOR LOSS AND RESONANCE CHARACTERISTICS

8.5.1. Decay Time and Q-Value

Here we discuss the decay time, τ_p, of the optical energy in a laser resonator and its corresponding resonance characteristics. The rate of change of the total number of photons in the resonator can be expressed as:

$$\frac{dS}{dt} = -\frac{1}{\tau_p} S \tag{8.105}$$

The solution of this equation can be readily written as:

$$S = S(t = 0)\exp(-t/\tau_p) \tag{8.106}$$

Here, one round-trip time in a resonator with a length, L, is $2nL/c$. We assume the propagation loss to be α_a in the medium and the mirror reflectivity to be R_1 and R_2 at either end. Since the decay after one round-trip should be equal to the total loss, we have:

$$e^{-2\alpha_a L}R_1R_2 = e^{-2nL/c\tau_p} \tag{8.107}$$

Thus,

$$\tau_p = \frac{nL/c}{\left(\alpha_a + \dfrac{1}{2L}\ln\dfrac{1}{R_1R_2}\right)L} = \frac{nL}{c\left(\alpha_a L + \dfrac{1}{2}\ln\dfrac{1}{R_1R_2}\right)} \tag{8.108}$$

Furthermore, the Q-value of the resonator is given by:

$$Q = \frac{\omega S}{P} = \frac{\omega S}{-dS/dt} \tag{8.109}$$

where S denotes the energy stored in the resonator and the electric power loss is given by $P = -dS/dt$. From the above expressions, we obtain[22]

$$Q = \omega\tau_p = \frac{k_0 nL}{\alpha_a L + \ln\sqrt{1/R_1R_2}} \tag{8.110}$$

where $k_0 = 2\pi/\lambda$.

8.5.2. Resonance Characteristics and Transfer Function

Figure 8.17 shows that a laser resonator can function as a filter. If gain is added to a laser medium, the filter bandwidth is decreased. There is a close relationship between the laser resonance characteristics, such as the Q-value, and a transfer function $H(\omega)$ of an equivalent filter. For example, the half-width, $\Delta\omega$, of the equivalent filter passband is written as:

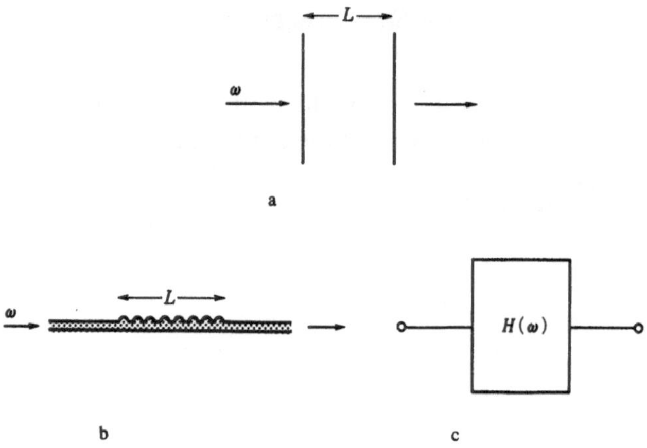

Figure 8.17. Resonators and filters. (a) Fabry–Perot-type filter; (b) DFB-type filter; (c) equivalent filter.

$$\Delta\omega = \frac{\omega}{Q} \tag{8.111}$$

8.6. Summary

Various types of laser resonators have been introduced in this chapter. Lasers using these resonators are widely applied not only in the optoelectronics area, but also in industrial and medical fields. The basic idea of a laser resonator was developed in the early 1960s, and the concept of a periodic structure resonator was considered in 1970. Laser resonators must be designed to meet the ever-more-demanding requirements for laser performance. This need is driving fundamental research and the development of materials and fabrication technology in order to produce high-performance laser resonator configurations. This chapter was intended to introduce fundamental concepts in laser resonator technology and to be helpful for the design of future resonator configurations.

PROBLEMS

8.1. Calculate the longitudinal mode separation $\Delta\lambda_l$ and transverse mode separation $\Delta\lambda_t$ of a resonator consisting of a distributed index waveguide by using Eqs. (8.15) and (8.16). Use numerical values listed in the example in the text except $\lambda = 1.3\ \mu$m.

8.2. Tabulate the relation between a small change of frequency Δf (GHz) and a small change of wavelength $\Delta\lambda$ (Å) for $\lambda = 0.78$, 1.3, and 1.55 μm.

8.3. In Eq. (8.12) we have neglected the dispersion effect of n_1. To include the dispersion in obtaining a longitudinal mode separation we must use an effective index n_{eff} defined by

$$n_{\text{eff}} = n_1[1 - (\lambda/n_1)(\partial n_1/\partial\lambda)]_{\lambda=\lambda_0}$$

Show this relationship.

8.4. Obtain the allowable z for a Fabry–Perot resonator when f and k are given. [Use Eq. (8.20).]

8.5. Obtain the wave-front coefficient at a concave mirror shown in Fig. 8.3b.

8.6. Obtain the spotsize s_0 of a confocal resonator for $f = 1$ m and $\lambda = 6328$ Å.

8.7. Derive Eq. (8.36).

8.8. Derive Eq. (8.59).

8.9. What is the degeneracy of resonant modes in a laser resonator?

8.10. In a DFB resonator we have assumed that $n_{\text{eq}}/n_2 = 1$. As a matter of fact, this term is larger than unity, since n_{eq} is the equivalent index of the core and n_2 is that of the cladding. Then Eq. (8.78) reads

$$\cos(\theta_m) = 1 - 2(n_{\text{eq}}/n_2)$$

for $N = 1$ and $m = 1$. Since $n_{\text{eq}}/n_2 > 1$, $\cos(\theta_m) < -1$. This means that we cannot find a suitable angle for Bragg diffraction and this statement is not true. Consider what is wrong in the above discussion.

8.11. Obtain eigenvalues of Eqs. (8.96) and (8.97). Hint: It is necessary to use a computer for a numerical calculation.

8.12. Obtain the width of the stop band of a DFB resonator. Express it in terms of the coupling coefficient κ.

8.13. Find the relation between the reflectivity and transmissivity of a laser resonator when we consider it as a filter.

REFERENCES

1. A. L. Schawlow and C. H. Townes, *Phys. Rev.* **112,** 1958 (1940).
2. T. H. Maiman, *Nature* **187,** 493 (1960).
3. A. G. Fox and T. Li, *Bell Syst. Tech. J.* **40,** 453 (1961).
4. G. D. Boyd and J. P. Gordon, *Bell Syst. Tech. J.* **40,** 489 (1961).
5. Y. Suematsu and T. Ikegami, *J. IECE Jpn.* **49,** 1091 (1966).
6. M. B. Panish, I. Hayashi, and S. Sumski, *Appl. Phys. Lett.* **16,** 326 (1970).
7. M. I. Nathan, W. P. Dumke, G. Burns, F. H. Dill, Jr., and G. Lasher, *Appl. Phys. Lett.* **1,** 62 (1962).
8. T. M. Quist, R. H. Rediker, R. J. Keyes, W. E. Krag, B. Lax, A. L. McWorter, and H. J. Zeigler, *Appl. Phys. Lett.* **1,** 91 (1962).

9. R. N. Hall, G. E. Fenner, J. D. Kingsley, T. J. Soltys, and R. O. Carlson, *Phys. Rev. Lett.* **9**, 366 (1962).
10. A. S. Dobkin, O. N. Korekov, G. A. Lapitskaya, A. A. Pleskikon, O. N. Prozorov, L. A. Rivlin, G. A. Sukhareva, V. S. Shildyaev, and S. D. Yakubovich, *Sov. Phys. Semicond.* **4**, 515 (1970).
11. K. Iga, T. Kanbayashi, K. Wakao, and Y. Sakamoto, *Jpn. J. Appl. Phys.* **18**, 2035 (1979).
12. M. Nakamura, A. Yariv, H. W. Yen, S. Somekh, and H. L. Garvin, *Appl. Phys. Lett.* **22**, 515 (1973).
13. S. Wang, *IEEE J. Quantum Electron.* **QE-10**, 413 (1970).
14. Y. Suematsu, M. Yamada, and K. Hayashi, *Proc. IEEE (Lett.)* **63**, 208 (1975).
15. L. A. Stratton, *Electromagnetic Theory,* McGraw–Hill, New York (1941).
16. Y. Suematsu and Y. Fukinuki, *J. IRCR Jpn.* **48**, 1684 (1965).
17. M. Born and E. Wolf, *Principles of Optics,* Pergamon Press, Elmsford, N.Y. (1975).
18. G. D. Boyd and H. Kogelnik, *Bell Syst. Tech. J.* **41**, 1347 (1962).
19. D. Marcuse, *Theory of Dielectric Waveguide,* Academic Press, New York (1974).
20. H. Kogelnik and C. V. Shank, *J. Appl. Phys.* **43**, 2327 (1972).
21. K. Utaka, S. Akiba, K. Sakai, and Y. Matsushita, *IEEE J. Quantum Electron.* **QE-22**, 1042 (July 1986).
22. A. Yariv, *Quantum Electronics,* 2nd ed., Wiley, New York (1975).

NINE

LASER EQUATIONS

The purpose of this chapter is to describe the density matrix formulation for the coupling of an electric field to a simple two-state atomic system, and to represent a dipole transition by using this approach. Through this formulation we can appreciate how the dipole transition occurs based on the optically driven interference between the energy states of an atom or molecule. Then, the so-called laser equations will be derived to obtain properties of laser oscillation and atomic–field interaction dynamics. Among other things, these equations will prove helpful in obtaining expressions for the saturation of laser gain. For a theoretical basis, we rely upon Pantell and Puthoff, cited in Ref. 1.

9.1. DENSITY MATRIX AND EQUATIONS OF MOTION

9.1.1. Density Matrix

Figure 9.1 shows two energy eigenstates of an electron bound to a quantum system. These states are expressed by Dirac's vector representation $|1\rangle$ and $|2\rangle$. The pure state $|\phi\rangle$ of the system is written by using the linear combination of eigenstates

$$|\phi\rangle = a|1\rangle + b|2\rangle \tag{9.1}$$

Here, $|a|^2$ and $|b|^2$ denote the probability that the electron is in each eigen-

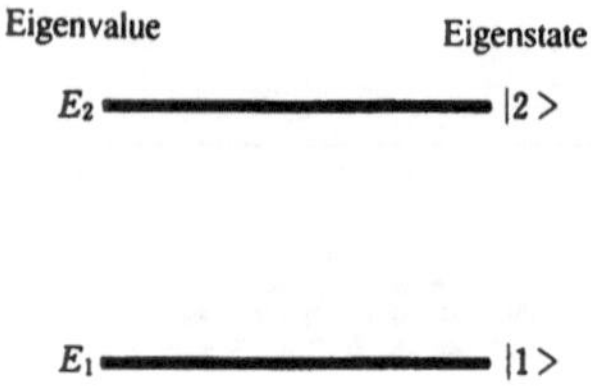

Figure 9.1. Eigenstates of two energies.

state of the superposition. In general, a mixed state is constructed according to the various combinations of superpositions as follows:

$$|\phi_1\rangle = a_1|1\rangle + b_1|2\rangle$$
$$|\phi_2\rangle = a_2|1\rangle + b_2|2\rangle \tag{9.2}$$

The expectation value of a physical observable A is written as:

$$\begin{aligned}\langle A\rangle &= \sum_n p_n\langle\phi_n|A|\phi_n\rangle \\ &= \sum_{n,m} p_n\langle\phi_n|A|m\rangle\langle m|\phi_n\rangle \\ &= \sum_{n,m} p_n\langle m|\phi_n\rangle\langle\phi_n|A|m\rangle \end{aligned} \tag{9.3}$$

where p_n is the probability of superposition state n and

$$\rho = \sum_n p_n|\phi_n\rangle\langle\phi_n| \tag{9.4}$$

where I is the identity (unit) matrix. Then, by defining the density operator

$$\langle A\rangle = \sum_m \langle m|\rho A|m\rangle = \mathrm{Tr}(\rho A) \tag{9.5}$$

the following equation is obtained from Eq. (9.4):

$$I = \sum_m |m\rangle\langle m| \tag{9.6}$$

where Tr( ) denotes the trace of the ρA matrix which is the diagonal sum.

9.1.2. Density Operator and Density Matrix in the Pure State

The density operator ρ in the pure state is defined as:

$$\begin{aligned}\rho &= |\phi\rangle\langle\phi| = (a|1\rangle + b|2\rangle)(a^*\langle 1| + b^*\langle 2|) \\ &= |a|^2|1\rangle\langle 1| + ab^*|1\rangle\langle 2| + a^*b|2\rangle\langle 1| + |b|^2|2\rangle\langle 2| \end{aligned} \tag{9.7}$$

The element ρ_{ij} of this density matrix operator is expressed as:

$$\rho_{ij} = \langle i|\rho|j\rangle \tag{9.8}$$

and is written in the pure state from Eq. (9.7) as:

$$(\rho) = \begin{pmatrix} \rho_{11} & \rho_{12} \\ \rho_{21} & \rho_{22} \end{pmatrix} = \begin{pmatrix} |a|^2 & ab^* \\ a^*b & |b|^2 \end{pmatrix} \tag{9.9}$$

The density operator ρ satisfies the following equation of motion:

$$j\hbar \frac{\partial \rho}{\partial t} = [\mathscr{H}, \rho] = \mathscr{H}\rho - \rho\mathscr{H} \tag{9.10}$$

where $\mathscr{H}$ is a Hamiltonian of the system. The above equation can be simplified to an equation of motion for the density matrix by separating the Hamiltonian:

$$\mathscr{H} = \mathscr{H}_0 + \mathscr{H}' \tag{9.11}$$

where $\mathscr{H}_0$ and $\mathscr{H}'$ are the unperturbed and perturbed components, respectively. The perturbed part denotes the interaction between light and the quantum system. Executing the operations $\langle i|$ and $|j\rangle$ on both sides of Eq. (9.10) gives:

$$j\hbar \frac{\partial \rho_{ij}}{\partial t} = (E_i - E_j)\rho_{ij} + [\mathscr{H}', \rho]_{ij} \tag{9.12}$$

where E_i and E_j are the eigenvalues i and j of the unperturbed Hamiltonian.

9.1.3. Density Operator in a Continuous Eigenstate

Density operators can also be defined in the case where the energy eigenstates of electrons in a semiconductor crystal are continuous. Figure 9.2 shows the eigenstate $|i, k\rangle$ which is characterized by the parameter k (in general, momentum is indicated by $\hbar k$). In this figure, $i = 1$ represents the lower energy band (valence band) and $i = 2$ represents the upper energy band (conduction band).

An arbitrary state $|\phi\rangle$ is expressed by the superposition of eigenstates as:

$$|\phi\rangle = \int d^3\mathbf{k}'_1 c_1(\mathbf{k}'_1)|1, \mathbf{k}'_1\rangle + \int d^3\mathbf{k}'_2 c_2(\mathbf{k}'_2)|2, \mathbf{k}'_2\rangle \qquad (9.13)$$

Here, there is an orthogonal relationship among eigenvectors:

$$\langle i, k_1 | j, k_2\rangle = \delta(k_1, k_2) \qquad (9.14)$$

where δ is the Kroneker delta function which equals 1 if $k_1 = k_2$ and is 0 otherwise.

The density operator of a continuous state is written in the same manner as that of the discrete one:

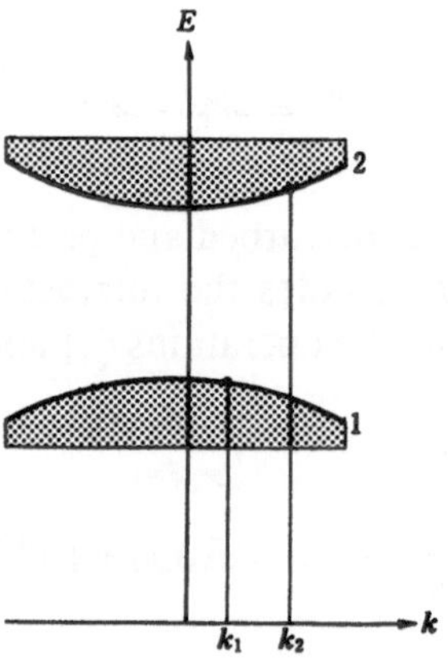

Figure 9.2. Continuous eigenstate.

$$\rho \equiv |\phi\rangle\langle\phi| = \iint d^3\mathbf{k}'_1 d^3\mathbf{k}'_2 c_1(\mathbf{k}'_1)c''_1(\mathbf{k}_2)|1, \mathbf{k}'_1\rangle\langle 1, \mathbf{k}'_2|$$
$$+ \iint d^3\mathbf{k}'_1 d^3\mathbf{k}'_2 c_1(\mathbf{k}'_1)c''_2(\mathbf{k}_2)|1, \mathbf{k}'_1\rangle\langle 2, \mathbf{k}'_2|$$
$$+ \iint d^3\mathbf{k}'_1 d^3\mathbf{k}'_2 c''_1(\mathbf{k}'_1)c_2(\mathbf{k}'_2)|2, \mathbf{k}'_2\rangle\langle 1, \mathbf{k}'_1|$$
$$+ \iint d^3\mathbf{k}'_1 d^3\mathbf{k}'_2 c_2(\mathbf{k}'_1)c''_2(\mathbf{k}'_2)|2, \mathbf{k}'_1\rangle\langle 2, \mathbf{k}'_2| \qquad (9.15)$$

The corresponding density matrix element is written as:

$$\rho_{ij}(\mathbf{k}_1, \mathbf{k}_2) = \langle i, \mathbf{k}_1|\rho|j, \mathbf{k}_2\rangle \qquad (9.16)$$

and becomes a function of k_1 and k_2. From the orthogonality of the eigenfunctions, we find:

$$\rho_{11}(\mathbf{k}_1, \mathbf{k}_2) = C_1(\mathbf{k}_1)C_1^*(\mathbf{k}_2)$$
$$\rho_{12}(\mathbf{k}_1, \mathbf{k}_2) = C_1(\mathbf{k}_1)C_2^*(\mathbf{k}_2)$$
$$\rho_{21}(\mathbf{k}_1, \mathbf{k}_2) = \rho^*_{12}(\mathbf{k}_1, \mathbf{k}_2)$$
$$\rho_{22}(\mathbf{k}_1, \mathbf{k}_2) = C_2(\mathbf{k}_1)C_2^*(\mathbf{k}_2) \qquad (9.16')$$

Commutation of the density matrix with the perturbed Hamiltonian, $\mathscr{H}'$, leads to:

$$[\mathscr{H}', \rho]_{ij} = \langle i, \mathbf{k}_1|[\mathscr{H}', \rho]|j, \mathbf{k}_2\rangle$$
$$= \langle i, \mathbf{k}_1|\mathscr{H}'\rho - \rho\mathscr{H}'|j, \mathbf{k}_2\rangle$$
$$= \sum_{l=i,j} \int \{\langle i, \mathbf{k}_1|\mathscr{H}'|l, \mathbf{k}'_1\rangle\langle l, \mathbf{k}'_1|\rho|j, \mathbf{k}_1\rangle$$
$$- \langle i, \mathbf{k}_1|\rho|l, \mathbf{k}'_1\rangle\langle l, \mathbf{k}'_1|\mathscr{H}'|j, \mathbf{k}_2\rangle\} d^3\mathbf{k}'_1$$
$$= \sum_{l=i,j} \int \{\mathscr{H}'_{il}(\mathbf{k}_1, \mathbf{k}'_1)\rho_{li}(\mathbf{k}'_1, \mathbf{k}_2)$$
$$- \rho_{il}(\mathbf{k}_1, \mathbf{k}'_1)\mathscr{H}'_{ij}(\mathbf{k}'_1, \mathbf{k}_2)\} d^3\mathbf{k}'_1 \qquad (9.17)$$

As before, the expectation value of the physical quantity, A, can be expressed as:

$$\begin{aligned} \langle A \rangle &= \langle \phi | A | \phi \rangle \\ &= \mathrm{Tr}(\rho A) \\ &= \sum_{ij} \iint \rho_{ij}(\mathbf{k}_1, \mathbf{k}_2) A_{ji} d^3\mathbf{k}_1 d^3\mathbf{k}_2 \end{aligned} \tag{9.18}$$

Moreover, the probability of finding the electron with momentum k is easily written as:

$$\rho_{ii}(\mathbf{k}) = \rho_{ii}(\mathbf{k}, \mathbf{k}) \tag{9.19}$$

Here, by detailing Eq. (9.16), we can obtain a matrix whose form has a double structure:

$$\begin{array}{cc} & \begin{array}{cccc} & \text{Zone 1} & & \text{Zone 2} \end{array} \\ \begin{array}{c} \text{Zone 1} \\ \\ \\ \text{Zone 2} \\ \\ \end{array} & \left(\begin{array}{cccc} & \mathbf{k}_1 \rightarrow & & \mathbf{k}_1 \rightarrow \\ \mathbf{k}_2 & & \downarrow & \\ \downarrow & & \mathbf{k}_2 & \\ & \mathbf{k}_1 \rightarrow & & \mathbf{k}_1 \rightarrow \\ \mathbf{k}_2 & & \downarrow & \\ \downarrow & & \mathbf{k}_2 & \end{array} \right) \end{array}$$

where $\rho_{11}(k_1, k_2)$ represents a matrix element for the different wave numbers k in Zone 1 and denotes the probability of finding an electron when $k_1 = k_2$ by the procedure introduced by Pantell and Puthoff.[1]

9.2. DIPOLE TRANSITION

In this section we explain how the interaction between the microscopic dipole transition moment, μ, and the optical field distribution, E, changes the energy of the system. The Hamiltonian representing the changed system energy is treated as a perturbation, that is:

$$\begin{aligned} \mathscr{H}' &= -\boldsymbol{\mu} \cdot \mathbf{E} \\ &= -(\mu_x E_x + \mu_y E_y + \mu_z E_z) \end{aligned} \tag{9.20}$$

9.2.1. Diagonal Elements of the Density Matrix

If, in Eq. (9.12), $i = j$, the first term becomes zero and the following equation is obtained:

$$j\hbar \frac{\partial \rho_{ii}}{\partial t} = [\mathscr{H}', \rho]_{ii}$$

$$= -\mathbf{E} \cdot \sum_k (\mu_{ik}\rho_{ki} - \rho_{ik}\mu_{ki}) \tag{9.21}$$

By approximation, the microscopic polarization originates from a dipole driven by the electric field so the diagonal element becomes zero and we can write:

$$(\mu) = \begin{pmatrix} 0 & \mu_{12} \\ \mu_{21} & 0 \end{pmatrix} \tag{9.22}$$

Consequently, Eq. (9.21) yields:

$$j\hbar \frac{\partial \rho_{11}}{\partial t} = -\mathbf{E} \cdot (\mu_{12}\rho_{21} - \rho_{12}\mu_{21})$$

$$j\hbar \frac{\partial \rho_{22}}{\partial t} = -\mathbf{E} \cdot (\mu_{21}\rho_{12} - \rho_{21}\mu_{12}) \tag{9.23}$$

From the above equations, the difference between the energy state densities is written as:

$$j\hbar \frac{\partial}{\partial t}(\rho_{22} - \rho_{11}) = -2\mathbf{E} \cdot (\mu_{21}\rho_{12} - \rho_{21}\mu_{12}) \tag{9.24}$$

Here, the relaxation between both states has been ignored. Assuming that the difference between the diagonal matrix elements of the states decays exponentially with respect to time, we can write:

$$\frac{\partial}{\partial t}(\rho_{22} - \rho_{11}) = -\frac{2}{j\hbar}\mathbf{E} \cdot (\mu_{21}\rho_{12} - \rho_{21}\mu_{12}) - \frac{(\rho_{22} - \rho_{11}) - \Delta\rho^{e}}{T_1} \tag{9.25}$$

where T_1 is the relaxation time and $\Delta\rho^{e}$ is the equilibrium state difference. The relaxation time, T_1, introduced phenomenologically at this point, is called a longitudinal relaxation time. The relaxation effect can be understood with the help of mode interaction effects[2] and can be mathematically expressed in terms of coupling to the zero point photon bath.

9.2.2. Nondiagonal Elements of the Density Matrix

Assuming that $i \neq j$ in Eq. (9.12), we can write:

$$j\hbar \frac{\partial \rho_{12}}{\partial t} = (E_1 - E_2)\rho_{12} + [\mathscr{H}', \rho]_{12}$$

$$j\hbar \frac{\partial \rho_{21}}{\partial t} = (E_2 - E_1)\rho_{21} + [\mathscr{H}', \rho]_{21} \tag{9.26}$$

From Eq. (9.22), the second term on the right-hand side becomes:

$$[\mathscr{H}', \rho]_{12} = -\mathbf{E}\cdot\mu_{12}(\rho_{22} - \rho_{11}) \tag{9.27}$$

Further, if we can write:

$$E_2 - E_1 = \hbar\omega_0 \tag{9.28}$$

then the following expressions are obtained:

$$j\hbar \frac{\partial \rho_{12}}{\partial t} = -\hbar\omega_0\rho_{12} - \mu_{12}\cdot\mathbf{E}(\rho_{22} - \rho_{11})$$

$$j\hbar \frac{\partial \rho_{21}}{\partial t} = +\hbar\omega_0\rho_{21} - \mu_{21}\cdot\mathbf{E}(\rho_{11} - \rho_{22}) \tag{9.29}$$

By introducing a transverse relaxation effect phenomenologically, and assigning T_2 to the relaxation time, Eq. (9.29) becomes:

$$\frac{\partial \rho_{12}}{\partial t} + \left(\frac{1}{T_2} - j\omega_0\right)\rho_{12} = -\frac{1}{j\hbar}\mu_{12}\cdot\mathbf{E}(\rho_{22} - \rho_{11})$$

$$\frac{\partial \rho_{21}}{\partial t} + \left(\frac{1}{T_2} - j\omega_0\right)\rho_{21} = +\frac{1}{j\hbar}\mu_{21}\cdot\mathbf{E}(\rho_{22} - \rho_{11}) \tag{9.30}$$

From this expression it is clear that ρ_{21} is equal to ρ_{12}^*.

The physically observable microscopic polarization is expressed as an expectation value:

$$\langle \mu \rangle = \mathrm{Tr}(\rho\mu)$$
$$= \mathrm{Tr}\begin{pmatrix} \rho_{11} & \rho_{12} \\ \rho_{21} & \rho_{22} \end{pmatrix}\begin{pmatrix} 0 & \mu_{12} \\ \mu_{21} & 0 \end{pmatrix} = \rho_{12}\mu_{21} + \rho_{21}\mu_{12} \tag{9.31}$$

Accordingly, the temporal change is:

$$\frac{\partial}{\partial t}\langle \mu \rangle = \dot{\rho}_{12}\mu_{21} + \dot{\rho}_{21}\mu_{12} \tag{9.32}$$

where the dot denotes the time derivative. Substitution of $\dot{\rho}_{12}$ and $\dot{\rho}_{21}$ from Eq. (9.30) yields:

$$\frac{\partial}{\partial t}\langle \mu \rangle = j\omega_0(\dot{\rho}_{12}\mu_{21} + \dot{\rho}_{21}\mu_{12}) - \frac{1}{T_2}\langle \mu \rangle \tag{9.33}$$

By differentiating Eq. (9.33) and substituting Eq. (9.30), we can obtain:

$$\frac{\partial^2}{\partial t^2}\langle \mu_\alpha \rangle + \frac{2}{T_2}\frac{\partial}{\partial t}\langle \mu_\alpha \rangle + \omega_0^2\langle \mu_\alpha \rangle = -\frac{2\omega_0}{\hbar}\mu_\alpha\mu_\beta(\rho_{22} - \rho_{11})E_\beta \tag{9.34}$$

where α and β denote the Cartesian components of μ and repeated indices imply a sum. The approximation of $\omega_0 \gg 1/T_2$ has also been used. If it is assumed that the polarizations are random, as is the case for a uniform gas, then:

$$|\mu_{12}|^2 \cong 3|\overline{\mu_{\alpha 12}}|^2 \tag{9.35}$$

On the average, a third of the $|\mu_{12}|^2$ contributes to each of the Cartesian components. Moreover, in the case where polarization occurs only in the direction of an electric field, the whole of $|\mu_{12}|^2$ contributes. Thus, $|\mu_{12}|^2$ represents the mean value and expression (9.34) may be rewritten as:

$$\frac{\partial^2}{\partial t^2}\langle \mu \rangle + \frac{2\partial}{T_2\partial t}\langle \mu \rangle + \omega_0^2\langle \mu \rangle = -\frac{2\omega_0}{\hbar}|\overline{\mu_{12}}|^2(\rho_{22} - \rho_{11})\mathbf{E} \tag{9.36}$$

The first term on the right-hand side of Eq. (9.25) is rewritten using Eq. (9.33) to give the expression:

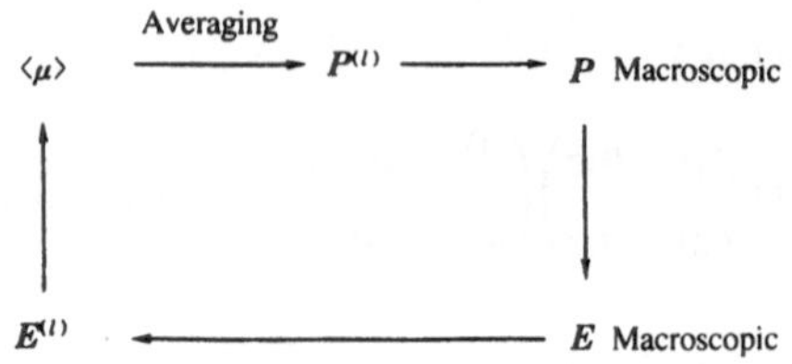

Figure 9.3. Microscopic and macroscopic polarization.

$$\frac{\partial}{\partial t}(\rho_{22}-\rho_{11})+\frac{1}{T_1}[(\rho_{22}-\rho_{11})-\Delta\rho^e]=\frac{2\mathbf{E}}{\hbar\omega_0}\langle\dot{\mu}\rangle \tag{9.37}$$

Equations (6.19), (9.34), and (9.37) are summarized:

$$\ddot{\mathbf{E}}+\frac{1}{\tau_p}\dot{\mathbf{E}}+\omega_c^2\mathbf{E}=-\frac{1}{\varepsilon}\ddot{\mathbf{P}} \tag{9.38}$$

$$\langle\ddot{\mu}\rangle+\frac{2}{T_2}\langle\dot{\mu}\rangle+\omega_0^2\langle\mu\rangle=-\frac{2\omega_0}{\hbar}|\overline{\mu_{12}}|^2(\rho_{22}-\rho_{11})\mathbf{E}^{(l)} \tag{9.38$'$}$$

$$(\rho_{22}-\rho_{11})+\frac{1}{T_1}[(\rho_{22}-\rho_{11})-\Delta\rho^e]=\frac{2E}{\hbar\omega_0}\cdot\langle\dot{\mu}\rangle\cdot\mathbf{E}^{(l)} \tag{9.38$''$}$$

where $\mathbf{E}^{(l)}$ is the local electric field and $\mathbf{E}$ is the macroscopic electric field. The above three equations are called the laser equations, and they determine the behavior of the light as well as the state of the molecules in a laser resonator.

The relation between the local and microscopic fields is written as:

$$\mathbf{P}=\frac{(n_r)^2+2}{3}\mathbf{P}^{(l)} \tag{9.39}$$

and

$$\mathbf{E}^{(l)}=\frac{(n_r^2)+2}{3}\mathbf{E} \tag{9.39$'$}$$

where n_r is the refractive index. These expressions lead to the Lorentz correction factor, L, which is:

$$L = [(n_r^2 + 2)/3]^2 \tag{9.40}$$

Here, $L = 1$v in a semiconductor, and $n_r = 1$ in a rare gas.

PROBLEMS

9.1. State the analogy between ket/bra vectors and spatial vectors.
9.2. Obtain Eq. (9.9) from Eq. (9.7).
9.3. By employing the Schrödinger equation, obtain the equation of motion for density operator Eq. (9.10).
9.4. Prove Eq. (9.12). Hint: Use the orthogonality of eigenvectors.
9.5. Explain why the perturbation Hamiltonian $\mathscr{H}'$ for a dipole transition is expressed by Eq. (9.20).
9.6. Prove Eq. (9.21).
9.7. Explain the physical meaning of a longitudinal relaxation time.
9.8. Obtain Eq. (9.27).
9.9. Explain the physical meaning of transverse relaxation times.
9.10. Show that the expectation value of dipole moment $\langle \mu \rangle$ is given by Eq. (9.31).
9.11. Obtain Eqs. (9.33) and (9.34).
9.12. Obtain Eqs. (9.36) and (9.37).
9.13. Explain how laser operation will be expressed by a series of laser equations (9.38)–(9.38″).
9.14. What is the Lorentz correction factor?

REFERENCES

1. R. H. Pantell and H. E. Puthoff, *Fundamentals of Quantum Electronics,* Wiley, New York (1969).
2. D. Marcuse, *Engineering Quantum Electrodynamics,* Harcourt, Brace & World, New York (1970).
3. A. Yariv, *Quantum Electronics,* 2nd ed., Wiley, New York (1975).
4. L. I. Schiff, *Quantum Mechanics,* McGraw-Hill, New York (1940).

TEN

RATE EQUATIONS

In this chapter, rate equations, which express the behavior of lasers, are derived from the laser equations introduced in Chapter 9. This discussion will enable us to determine the oscillation conditions of lasers and to understand how the pump power is transferred into a laser beam. The concept of optical gain is introduced for a two-level system and for semiconductors. It is also shown that the gain of a quantum well structure behaves like that of a two-level system. In the last section we will introduce some solutions of rate equations. The analytical treatments in 10.1 and 10.2 is based on Pantell and Puthoff, cited in Ref. 1.

10.1. HOMOGENEOUS GAIN

Quantum systems (atoms, molecules, electrons, etc.) which lead to light amplification have atomic-scale dimensions. The wavelength of light is, on the other hand, 10^2 to 10^3 times larger than the size of these atoms and molecules. The ratio of light wavelength to the de Broglie wavelength of electrons in semiconductor crystals is on the same order. Therefore, it is reasonable to assume that the change of the optical field is gradual from an atomic point of view.

When a group of quantum systems have the same or almost the same interaction with an electric field of light, the group is said to be homogeneous. On the other hand, when a group of quantum systems interacts independently, it is said to be inhomogeneous. For example, the energy levels of ions in solids are all similarly affected by a crystalline field, so the energy broadening relating to the crystal potential is homogeneous. However, atoms and molecules in a gaseous state move randomly in space and each sees a different local field because of the Doppler effect. The energy broadening is, therefore, inhomogeneous. The behavior of electrons in a semiconductor may be considered to be nearly homogeneous, but in the case where the interaction is influenced by a relaxation time (10^{-14} to 10^{-13} sec) in a conduction band, the behavior is considered to be partially inhomogeneous.

In a quantum system having homogeneous gain, obtaining the macroscopic polarization, P, from the microscopic polarization, $\langle \mu \rangle$, is rather easy, because all the microscopic polarizations have the same effect, so the total effect is the sum of microscopic polarizations in a given volume. Now by assuming the microscopic local polarization of the ith molecule to be $\langle \mu \rangle_i$, the macroscopic local polarization can be expressed as a mean value $\mathbf{P}^{(l)}$:

$$\mathbf{P}^{(l)} = \frac{1}{V} \sum_{i=1}^{M} \langle \mu \rangle_i = \frac{M}{V} \langle \bar{\mu} \rangle \tag{10.1}$$

where V, M, and $\langle \bar{\mu} \rangle$ denote the volume, total, and mean values, respectively (Fig. 10.1). The superscript (l) indicates the spatially local quantity.

The state density is obtained from the diagonal components of the density matrix:

$$\frac{1}{V} \sum_{i=1}^{M} (\rho_{22} - \rho_{11})_i = \frac{M}{V} (\overline{\rho_{22} - \rho_{11}}) \equiv N_2 - N_1 \tag{10.2}$$

where N_1 and N_2 are the number of oscillators per unit volume in state 1 and state 2, and are called the population of the state.

Averaging Eqs. (9.38′) and (9.38″) yields:

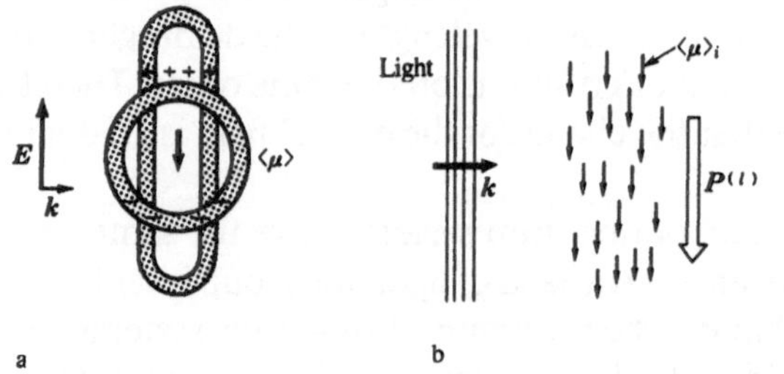

Figure 10.1. (a) Microscopic and (b) macroscopic polarization.

$$\ddot{\mathbf{P}}^{(l)} + \frac{2}{T_2}\dot{\mathbf{P}}^{(l)} + \omega_0^2\mathbf{P}^{(l)} = -\frac{2\omega_0}{\hbar}\,|\overline{\mu_{12}}|^2(N_2 - N_1)\cdot\mathbf{E}^{(l)} \tag{10.3}$$

$$(\dot{N_2 - N_1}) + \frac{1}{T_1}[(N_2 - N_1) - N^e] = \frac{2}{\hbar\omega_0}\dot{\mathbf{P}}^{(l)}\cdot\mathbf{E}^{(l)} \tag{10.4}$$

From the Lorentz correction factor we can write:

$$\mathbf{P} = \sqrt{\mathcal{L}}\mathbf{P}^{(l)}, \qquad \mathbf{E}^{(l)} = \sqrt{\mathcal{L}}\mathbf{E} \tag{10.5}$$

Then, by rewriting the above equations and including Eq. (9.38), the following equations are obtained:

$$\ddot{\mathbf{E}} + \frac{1}{\tau_p}\dot{\mathbf{E}} + \omega_c^2\mathbf{E} = -\frac{1}{\varepsilon}\ddot{\mathbf{P}} \tag{10.6}$$

$$\ddot{\mathbf{P}} + \frac{2}{T_2}\dot{\mathbf{P}} + \omega_0^2\mathbf{P}^{(\iota)} = -\frac{2\omega_0}{\hbar}\,|\overline{\mu_{12}}|^2\mathcal{L}(N_2 - N_1)\cdot\mathbf{E}$$

$$(\dot{N_2 - N_1}) + \frac{1}{T_1}[(N_2 - N_1) - N^e] = \frac{2}{\hbar\omega_0}\dot{\mathbf{P}}\cdot\mathbf{E} \tag{10.6$'$}$$

where $N^e = \Delta\rho^e \times (M/V)$, with M and V defined after Eq. (10.1):

$$\Delta\rho^e = (\rho_{22}^e - \rho_{11}^e) \tag{10.6$''$}$$

10.2. RATE EQUATIONS

Equations (10.6), (10.6′), and (10.6″) are said to be laser equations and describe the dynamic behavior of lasers. These expressions are sometimes inconvenient to handle since they are interdependent. Let us therefore derive approximate equations in order to simply express the dynamic response of lasers. Here we assume that optical power changes slowly compared with the frequency of light, which is a good approximation since optical frequencies are on the order of 10^{14} to 10^{15} Hz.

The polarization, $\mathbf{P}$, and the optical field, $\mathbf{E}$, are written as:

$$\mathbf{P} = \tfrac{1}{2}\,\mathbf{p}e^{j\omega t} + \text{c.c.} \tag{10.7}$$

$$\mathbf{E} = \tfrac{1}{2}\mathbf{e}e^{j\omega t} + \text{c.c.} \tag{10.8}$$

where c.c. is the complex conjugate. The time derivatives of the polarization, **P**, then become:

$$\dot{\mathbf{P}} = \tfrac{1}{2}[\dot{\mathbf{p}}e^{j\omega t} + \mathbf{p}j\omega e^{j\omega t} + \text{c.c.}] \tag{10.9}$$

$$\ddot{\mathbf{P}} = \tfrac{1}{2}[\ddot{\mathbf{p}}e^{j\omega t} + 2\dot{\mathbf{p}}(j\omega)e^{j\omega t} - \omega^2\mathbf{p}e^{j\omega t} + \text{c.c.}] \tag{10.10}$$

Assuming that angular frequency variation of the polarization envelope (angular frequency) is s, then:

$$\dot{\mathbf{p}} \simeq js\mathbf{p} \tag{10.11}$$

When $\omega \gg |s|$, the quadratic differential term with respect to time, $\mathbf{p}$, can be omitted, and when $\omega \gg 1/T_2$, Eq. (10.6′) becomes:

$$\dot{\mathbf{p}} + \frac{1}{T_2}\mathbf{p} \simeq \left(\frac{j}{\hbar}\right)\mathcal{L}|\overline{\mu_{12}}|^2 N\cdot\mathbf{e}\left(\frac{\omega_0}{\omega}\right) \tag{10.12}$$

where $N = N_2 - N_1$.

The same approximations lead to a simplification of the population equation (10.6″):

$$\dot{N} + \frac{1}{T_1}(N - N^e) = \frac{1}{2}\left(\frac{j}{\hbar}\right)[\mathbf{p}\cdot\mathbf{e}^* - \mathbf{p}^*\cdot\mathbf{e}] \tag{10.13}$$

where * denotes the complex conjugate. The electric field equation (10.6) also simplifies to give:

$$\dot{\mathbf{e}} + \frac{1}{2\tau_p}\mathbf{e} = -j\left(\frac{\omega}{2\varepsilon}\right)\mathbf{p} \tag{10.14}$$

where $\omega = \omega_c$.

The solution of the polarization equation (10.12) is written as:

$$\mathbf{p} = e^{-t/T_2}\int^t e^{t'/T_2}\left(\frac{j}{\hbar}\right)\left(\frac{\omega_0}{\omega}\right)\mathcal{L}|\overline{\mu_{12}}|^2 N\cdot\mathbf{e}\,dt' \tag{10.15}$$

The relaxation time T_2 of the polarization is assumed to be much shorter than that of light in a resonator ($T_2 \ll \tau_p$). This assumption holds true in the case of gas or solid-state lasers. In the case of semiconductor lasers, $T_2 = 10^{-13}$ to 10^{-14} sec and $\tau_p = 10^{-12}$ sec, but the assumption is generally also applied to this case. With this assumption, all but the $\exp(t'/T_2)$ term can be removed from the integral, and the integration leads to:

$$\mathbf{p} \simeq \frac{jT_2}{\hbar} \cdot \left(\frac{\omega_0}{\omega}\right) \mathcal{L} |\overline{\mu_{12}}|^2 N \cdot \mathbf{e} \tag{10.16}$$

where $N_2 - N_1 = N$.

From Eq. (10.16) it is clear that a polarization with a phase shift $j = e^{j(\pi/2)}$ is induced by the optical field e.

Substituting Eq. (10.16) into (10.14) yields:

$$\dot{\mathbf{e}} + \frac{1}{2\tau_p}\mathbf{e} = -j\left(\frac{\omega}{2\varepsilon}\right)\frac{jT_2}{\hbar}\mathcal{L} \cdot |\overline{\mu_{12}}|^2 N \cdot \mathbf{e}\left(\frac{\omega_0}{\omega}\right)$$

$$= \frac{\omega_0 T_2}{2\hbar\varepsilon}\mathcal{L}|\overline{\mu_{12}}|^2 N \cdot \mathbf{e} \tag{10.17}$$

Taking the inner product of both sides of this equation with $\mathbf{e}^*$ leads to:

$$|\dot{\mathbf{e}}|^2 + \frac{1}{\tau_p}|\mathbf{e}|^2 = \frac{\omega_0 T_2}{\hbar\varepsilon}\mathcal{L}|\overline{\mu_{12}}|^2 N |\mathbf{e}|^2 \tag{10.18}$$

The time-averaged energy is expressed as $(1/2)\varepsilon|\mathbf{e}|^2$ where ε is the dielectric constant of the medium. The number of photons in the cavity when field quantization is considered then becomes:

$$S = \tfrac{1}{2}\varepsilon|\mathbf{e}|^2/\hbar\omega \tag{10.19}$$

Rewriting Eq. (10.18) using the above expression gives:

$$\dot{S} + \frac{1}{\tau_p}S = GNS \tag{10.20}$$

where the gain coefficient, G, is defined as:

$$G \simeq \frac{\omega_0 T_2}{\hbar\epsilon}\mathcal{L}|\overline{\mu_{12}}|^2 \tag{10.21}$$

From the approximation, $\dot{\mathbf{P}} = j\omega p$, and using Eq. (10.16), Eq. (10.6″) is transformed to:

$$\dot{N}+\frac{1}{T_1}(N-N^e)=-2GNS \tag{10.22}$$

Here, Eqs. (10.20) and (10.22) are summarized for reference:

$$\dot{S}+\frac{1}{\tau_p}S=GNS \tag{10.20$'$}$$

$$\dot{N}+\frac{1}{T_1}(N-N^e)=-2GNS \tag{10.22$'$}$$

The above pair of nonlinear equations are called rate equations. The rate of variation of the number of photons is given by the product of the population and the number of photons, S, in the first equation, which denotes induced emission. The second equation shows that the population is affected by induced emission. Here, if one photon is produced by a stimulated emission, the difference, N, of the population changes by two, which is reflected by the factor of 2 on the right-hand side of Eq. (10.22).

10.3. LASER GAIN

As shown in Sections 10.1 and 10.2, the incident electromagnetic wave induces a polarization in the medium, which oscillates at the optical frequency (Fig. 10.2). In the case where the phase of the oscillation of the polarization lags behind the phase of the incident electromagnetic wave, energy from the incident electromagnetic wave is absorbed by the polarization and the wave suffers a loss. In the case where the phase of the polarization is advanced with respect to the electromagnetic wave, the wave experiences gain and is amplified.

In Sections 10.1 and 10.2, we derived equations for the polarization and the electric field based on a density matrix formulation. From this we obtained the rate equation using a rotating wave ($e^{j\omega t}$) approximation. This

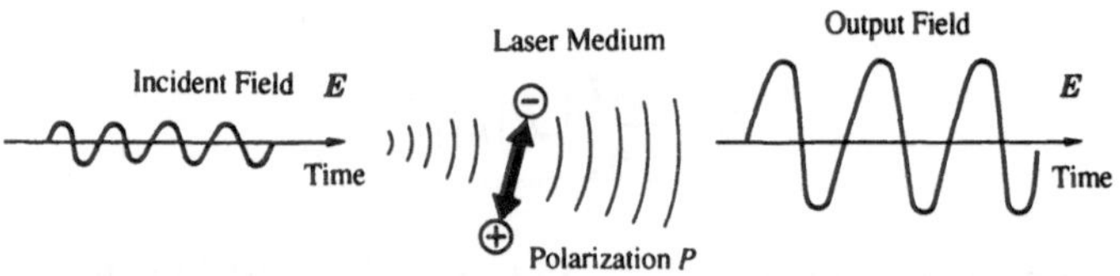

Figure 10.2. Polarization of a laser medium induced by incident electromagnetic waves.

section describes the gain which is experienced by the electromagnetic wave and which leads to an oscillation condition for lasers. In addition, we discuss here the gain of a semiconductor laser which contains a continuum of eigenstates.

10.3.1. Laser Gain

From Eq. (10.20), a rate equation denoting the temporal change of photons is written as:

$$\dot{S} = \left(GN - \frac{1}{\tau_p} \right) S \tag{10.23}$$

where

$$GN = \frac{\omega_0 T_2}{\hbar \varepsilon} \mathcal{L} |\overline{\mu_{12}}|^2 N \tag{10.24}$$

If the first term on the right-hand side of Eq. (10.23) is positive, which occurs when

$$N = N_2 - N_1 > 0 \tag{10.25}$$

then this term denotes the positive increment ratio (i.e., the amplification ratio) of a photon per unit time. The factor GN is called a gain factor. Equation (10.25) shows that the population density of the upper level must be larger than that of the lower level for the gain factor to be positive. This condition is called population inversion. One of the ways to achieve a population inversion is by the excitation of higher energy levels by means of an electric discharge in a gas laser. Other examples include optical pumping by a flash lamp which is used in solid-state lasers, and current injection into the conduction band which is used in semiconductor lasers. Increasing the population difference N so as to cause population inversion is therefore called excitation or pumping.

When the gain factor, GN, of the first term on the right-hand side of Eq. (10.23) exceeds the less factor $1/\tau_p$ in the second term, the number of photons, S, is amplified as a function of time and starts to increase drastically. This means that the laser begins to oscillate. The condition wherein the gain factor equals the loss factor is called the threshold (or oscillation) condition and is written as:

$$GN = \frac{1}{\tau_p} \tag{10.26}$$

or

$$N = \frac{1}{G\tau_p} \simeq N_{\text{th}} \tag{10.26'}$$

where N_{th} is called the threshold population difference.

When the population is pumped up to a value larger than the threshold, S seems to increase to infinity with time [from Eq. (10.26)]. However, saturation of the gain occurs at high powers and S remains finite.

In Section 10.4, the oscillation conditions and stationary solutions are discussed through rate equations on the basis of the concept stated above. The gain factor in Eq. (10.24) is derived for ω (angular frequency of the incident electromagnetic wave) = ω_0 (characteristic angular frequency of polarization, i.e., the energy difference/$\hbar$ between the two levels). If $\omega \neq \omega_0$, Eq. (10.24) becomes:

$$GN = \frac{\omega_0}{\hbar\varepsilon} \frac{1/T_2}{(\omega - \omega_0)^2 + 1/T_2^2} |\overline{\mu_{12}}|^2 \mathcal{L}N \tag{10.27}$$

and has a Lorentzian spectral characteristic. This equation will be referred to later when semiconductors are discussed.

10.3.2. Gain of Semiconductor Lasers

Here we expand the above discussion into one for multilevel systems, i.e., a semiconductor. It is well known that the energy levels of semiconductors form band structures (see Fig. 10.3). The induced emission of light oc-

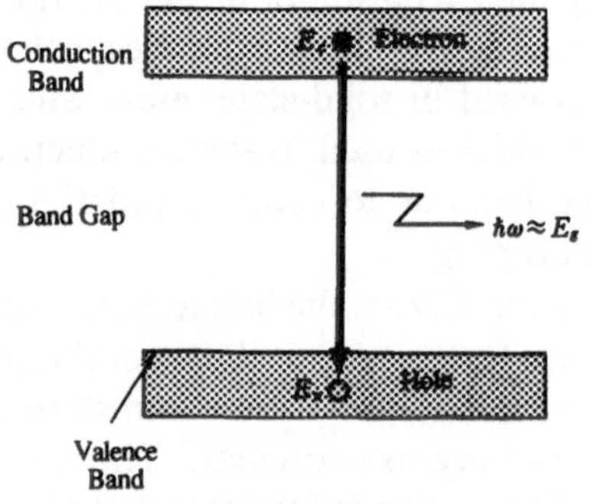

Figure 10.3. Band structure of a semiconductor and emission of light.

curs from the recombination between electrons in a conduction band and holes in a valence band (in other words, an electron–hole pair forms a polarization). In order to extend Eq. (10.27) to a semiconductor, it is necessary to add the band effects and to consider Pauli's exclusion principle with respect to the population, N. When Pauli's exclusion principle is taken into consideration, the probability of recombination between electrons in the conduction band and holes in the valence band is in proportion to $f_c(1-f_v)$. Here, f_c and f_v are Fermi functions for electrons and holes (which will be described later). In the same manner, the probability of absorption of light by electrons in the conduction band and holes in the valence band is in proportion to $(1-f_c)f_v$. Therefore, the net probability of emission between electrons and holes is given as:

$$f_c(1-f_v)-f_v(1-f_c)=f_c-f_v \tag{10.28}$$

The population N is written by multiplying f_c-f_v by the state density, g_{cv}:

$$N=g_{cv}(f_c-f_v) \tag{10.29}$$

The Fermi functions and the state density function are written as follows (see Fig. 10.4):

$$f_c=1\bigg/\left[1+\exp\left(\frac{E_c-E_{fc}}{kT}\right)\right] \tag{10.30}$$

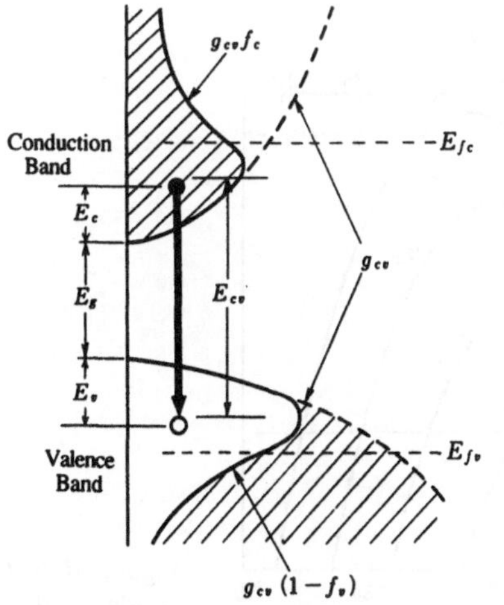

Figure 10.4. Energy population of electrons in a semiconductor.

$$f_v = 1 \Big/ \left[1 + \exp\left(\frac{E_v - E_{fv}}{kT}\right)\right] \qquad (10.30')$$

$$g_{cv} = \frac{1}{2\pi^2}\left(\frac{2m^*}{\hbar^2}\right)^{3/2}\sqrt{E_c + E_v - E_g} \qquad (10.31)$$

$$m^* = m_c^* m_v^* / (m_c^* + m_v^*) \qquad (10.31')$$

where m_c, m_v and E_{fc}, E_{fv} are effective mass and quasi-Fermi levels of electrons and holes, respectively. E_g is the energy gap between the conduction and the valence bands. By substituting Eq. (10.29) into (10.27), the gain factor of a semiconductor laser is given by:

$$G \equiv \int GNdE_{cv} = \int_{E_g}^{\infty} |\bar{\mu}_{12}|^2 \frac{\omega_0/(\hbar\varepsilon)\cdot 1/T_2}{(\omega - \omega_0)^2 + 1/T_2^2} g_{cv}(f_c - f_v)dE_{cv} \qquad (10.32)$$

where $E_{cv} = E_c - E_v$.

Figure 10.5 shows an example of a gain spectrum for various injected electron densities.[2] It is clear that the gain becomes positive with the increase of injected electron densities. From Eq. (10.32), it is also clear that, from the inverted population condition, f_c must be larger than f_v. This condition is given by Eq. (10.30) and leads to the expression:

$$E_{fc} - E_{fv} > E_{cv} - E_g \qquad (10.33)$$

10.3.3. Quantum Well Lasers

Let us consider a potential well as shown in Fig. 10.6 which is formed by means of a heterojunction of two kinds of semiconductors with different

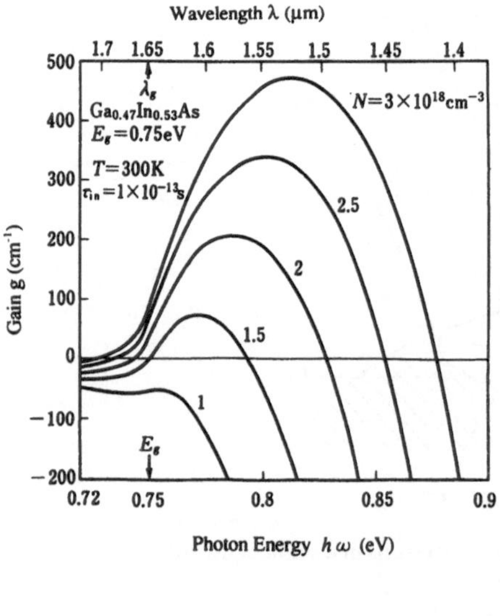

Figure 10.5. Gain spectrum. [After M. Asada and Y. Suematsu, *IEEE J. Quantum Electron.* **QE-21**, 434 (1985).]

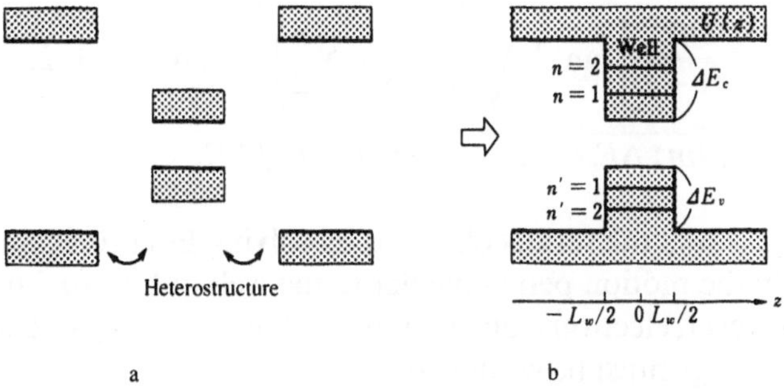

Figure 10.6. Quantum well laser. (a) Before and (b) after jointing.

band gaps. When the electrons are put into the well, the confinement causes them to possess discrete energy levels within the well, as is discussed in textbooks on elementary quantum mechanics. In other words, the continuous energy band becomes discrete in the direction in which the well-type potential is formed. This is a quantum size effect by which the energy eigenstates are modified by the artificial potential. A semiconductor laser which uses a quantum well, as shown in Fig. 10.6, to generate the lasing transitions, is called a quantum well laser.

Motion of conduction band electrons in the well is expressed by the following Schrödinger equation:

$$\left[-\frac{\hbar^2}{2m_c^*}\frac{\partial^2}{\partial z^2}+U(z)\right]\psi(z)=E\psi(z) \qquad (10.34)$$

where $U(z)$ is the well-type potential, and E is an eigenvalue of energy which will be obtained later. The solution of Eq. (10.34) becomes:

$$\psi(z)=\begin{cases} A\begin{pmatrix}\cos\\ \sin\end{pmatrix}[\kappa z] & (|z|\le L_w/2)\\ A\begin{pmatrix}\cos\\ \frac{z}{|z|}\sin\end{pmatrix}[\kappa L_w/2]e^{-\gamma(|z|-L_w/2)} & (|z|\ge L_w/2)\end{cases} \qquad (10.35)$$

where $\kappa = \sqrt{2m_c^* E}/\hbar$, $\gamma = \sqrt{2m_c^*(\Delta E_c - E)}/\hbar$. From the condition that a wave function and its first-order derivative are continuous at the boundary of a well, the following equation can be derived and used as a characteristic function to find the energy, E_n (compare this with the analysis for an optical waveguide in Chapter 7):

$$V = \frac{1}{\sqrt{1-b}}\left[\tan^{-1}\sqrt{\frac{b}{1-b}} + \bar{N}\frac{\pi}{2}\right] \qquad (\bar{N} = 0, 1, 2, \ldots) \quad (10.36)$$

$$V = \sqrt{2m_c^* \Delta E_c} L_w/\hbar, \qquad b = 1 - E_{\bar{N}}/\Delta E_c \quad (10.36')$$

Figure 10.7 shows the results obtained by solving Eq. (10.36) numerically. So far, only the motion perpendicular to the well in Fig. 10.6 has been discussed. However, electrons can also move in the direction parallel to the well, and so the energy must be written as:

$$E = E_n + \frac{\hbar^2 k_\parallel^2}{2m^*} \quad (10.37)$$

where $\hbar k_\parallel$ is a momentum parallel to the junction of the well. This term in the above equation is just the kinetic energy of electrons moving parallel to the well.

Because energy states become discrete, the state density of electrons is written (see Fig. 10.8) as:

$$g_{cv} = \sum_{n=1}^{\infty} \frac{1}{2\pi L_w}\left(\frac{2m^*}{\hbar^2}\right) H(E - E_n) \quad (10.38)$$

where $H(E - E_n)$ is the Heaviside step function.[4]

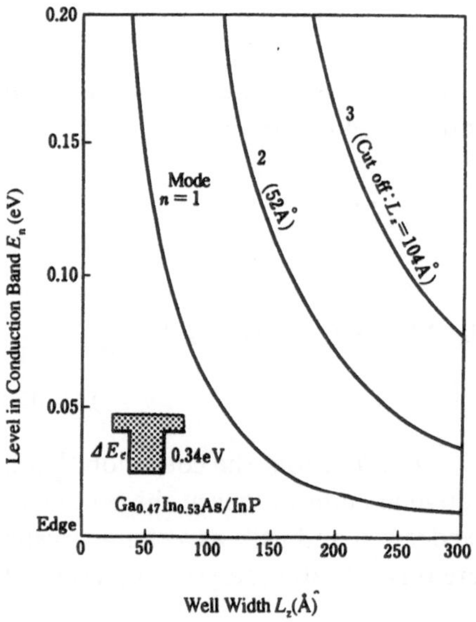

Figure 10.7. Quantization level. [After M. Asada, A. Kameyama, and Y. Suematsu, *IEEE J. Quantum Electron.* **QE-20,** 745 (1984).]

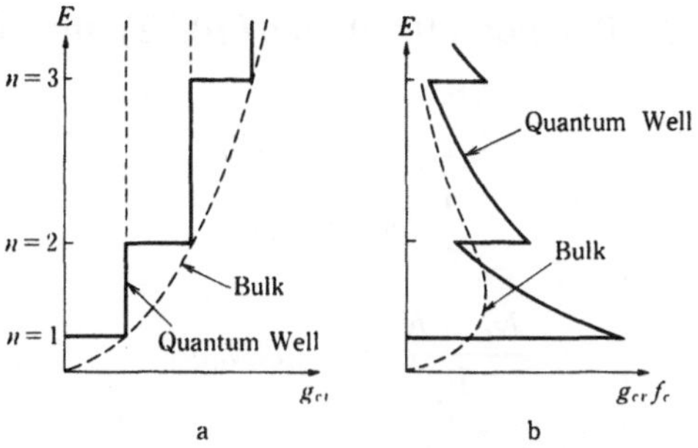

Figure 10.8. (a) State density $g_{cv}(E)$. (b) Carrier distribution $g_{cv}(E)f_c(E)$.

Note that the state density becomes stepwise and the spread of electrons becomes narrow. Consequently, as is seen from Fig. 10.9, gain becomes larger and sharper than that of ordinary bulk semiconductors.[3] The quantum well laser features an ultralow threshold, the control of oscillation wavelength by the design of the well thickness, an excellent temperature characteristic, and a narrow spectral gain width. In practice, a device with a threshold lower than 1 mA has been realized.

10.4. OSCILLATION CONDITIONS

In this section, we first discuss the stationary oscillation solution, and then introduce time-varying effects which lead to relaxation oscillation. By

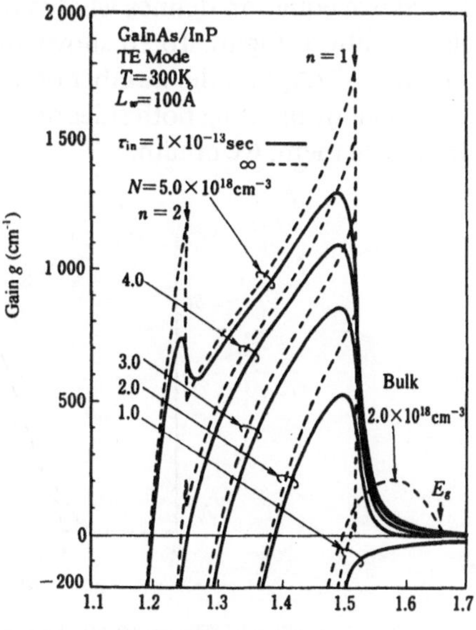

Figure 10.9. Gain spectrum of a quantum well laser. τ_{in} denotes the relaxation time in a band and nearly corresponds to T_2. [After M. Asada, A. Kameyama, and Y. Suematsu, *IEEE J. Quantum Electron.* **QE-20**, 745 (1984).]

assuming that $d/dt = 0$ in Eqs. (10.20) and (10.22), the solutions S_0 and N_0 are written as:

$$\frac{1}{\tau_p} S_0 = G N_0 S_0 \tag{10.39}$$

$$\frac{N_0 - N^e}{T_1} = -2 G N_0 S_0 \tag{10.40}$$

Here, if $S_0 \neq 0$, then:

$$N_0 = \frac{1}{G\tau_p} \tag{10.41}$$

$$S_0 = \frac{\tau_p}{2T_1} N_0 \left(\frac{N^e}{N_0} - 1 \right) \tag{10.42}$$

In order to achieve oscillation, $S_0 \geq 0$. Accordingly, these two equations lead to the following expression:

$$N^e \geq N_0 = 1/(G\tau_p) \tag{10.43}$$

The above equation defines what is required to reach the oscillation threshold condition. Figure 10.10 shows the relationship between the excitation quantity N^e/N_0 and the number of photons.

Next, by dividing both sides of Eqs. (10.20) and (10.22) by S_0 and N_0 to normalize them, we obtain:

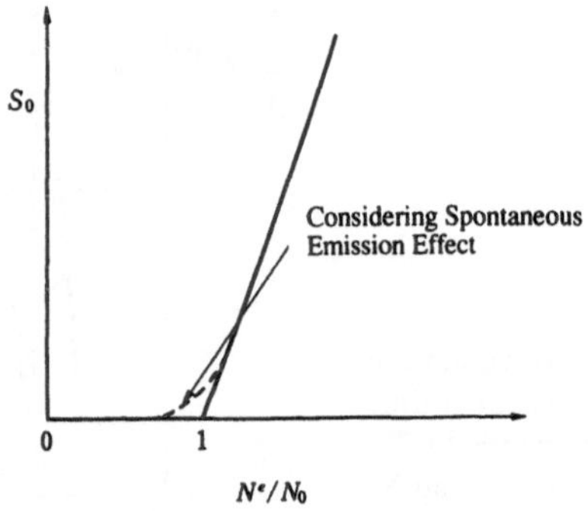

Figure 10.10. Excitation and light output.

$$\left(\frac{\dot{N}}{N_0}\right)+\frac{1}{T_1}\left(\frac{N}{N_0}-\frac{N^e}{N_0}\right)=-2\frac{N}{N_0}\frac{S}{S_0}GS_0 \tag{10.44}$$

$$\left(\frac{\dot{S}}{S_0}\right)+\frac{1}{\tau_p}\frac{S}{S_0}=GN_0\frac{S}{S_0}\frac{N}{N_0} \tag{10.45}$$

If $S/S_0=\bar{S}$, $N/N_0=\bar{N}$, and $N^e/N_0=\bar{N}^e$, the above two equations become:

$$\dot{N}+\frac{1}{T_1}(\bar{N}-\bar{N}^e)=-\frac{1}{T_1}\bar{N}\bar{S}(\bar{N}^e-1) \tag{10.46}$$

$$\dot{\bar{S}}+\frac{1}{\tau_p}\bar{S}=\frac{1}{\tau_p}\bar{N}\bar{S} \tag{10.47}$$

Then, from the condition that photons are excited stationarily by S_0, the effects of a slight change can be calculated by means of a perturbation (small signal) analysis. The following equations are derived by linearization of the nonlinear equation:

$$h\cdot\dot{q}+\frac{1}{T_1}(1+h\cdot q-\bar{N}^e)=-\frac{1}{T_1}(1+h\cdot q)(1+h\cdot p)(\bar{N}^e-1) \tag{10.48}$$

$$h\cdot\dot{p}+1/\tau_p(1+hp)=1/\tau_p(1+hq)(1+hp) \tag{10.48$'$}$$

where

$$S=1+h\cdot p(t) \tag{10.49}$$

$$N=1+h\cdot q(t) \tag{10.49$'$}$$

If $h \ll 1$ and we neglect the higher-order small terms, the following equations are obtained:

$$\dot{p}=\frac{1}{\tau_p}q \tag{10.50}$$

$$\dot{q}+\frac{\bar{N}^e}{T_1}q=\frac{1}{T_1}(1-\bar{N}^e)p \tag{10.51}$$

By taking the second derivative of Eq. (10.50) and substituting Eq. (10.51), the self-oscillation equation can be written:

$$\ddot{p} + \frac{\bar{N}^e}{T_1}\dot{p} + (\bar{N}^e - 1)\frac{1}{T_1\tau_p}p = 0 \tag{10.52}$$

One of the solutions of this linearized equation leads to relaxation oscillations and it may be written as:

$$p = e^{-\sigma t}[A\cos\omega_r t + B\sin\omega_r t] \tag{10.53}$$

where

$$\sigma = \frac{\bar{N}^e}{2T_1} \tag{10.54}$$

$$\omega_r \simeq \frac{1}{T_1}\sqrt{\frac{T_1}{\tau_p}(\bar{N}^e - 1) - \bar{N}^{e^2}/4} \tag{10.55}$$

Here, if $T_1/\tau_p \gg 1$, then the relaxation oscillation frequency, $f_r = \omega_r/2\pi$, is written as:

$$f_r = \frac{1}{2\pi T_1}\sqrt{\frac{T_1}{\tau_p}(\bar{N}^e - 1)} \tag{10.56}$$

PROBLEMS

10.1. What is the size of the wave packet of an electron with minimum uncertainty? Also, explain the de Broglie wavelength. Hint: Use a textbook on quantum mechanics.[6]

10.2. Derive Eqs. (10.6)–(10.6″).

10.3. Obtain Eqs. (10.12)–(10.14).

10.4. Obtain Eq. (10.20).

10.5. Obtain Eq. (10.22).

10.6. Try to obtain Eq. (10.27) when a frequency offset $\omega - \omega_0$ exists. Hint: In Eq. (10.16) we have approximated the solution. You must solve Eq. (10.12) including the frequency offset.

10.7. What is the state density in semiconductors?

10.8. Obtain an analytic formula Eq. (10.36) which expresses the energy eigenvalue of a quantum well. Hint: Compare Eq. (10.36) and the result in Chapter 3 on a planar waveguide. Use the relationship

$$(\kappa L_w/2)^2 + (rL_w/2)^2 = V^2$$

10.9. Obtain Eq. (10.37). Refer to Ref. 5.

10.10. Try to obtain the state density of a quantum well which is given by Eq. (10.38) and illustrated by Fig. 10.8a.

10.11. In Eq. (10.38), show that g_{cv} approaches Eq. (10.31) when $L_w \rightarrow \infty$.

10.12. What is a quantum well?

10.13. Derive Eqs. (10.45) and (10.46).

10.14. Obtain Eqs. (10.53) and (10.54).

10.15. Consider how we can include spontaneous emission in the rate equations. Hint: The spontaneous emission in this case is the factor which expresses the contribution of photons coming from the term $(N - N^e)/T_1$ in Eq. (10.22).

REFERENCES

1. R. H. Pantell and H. E. Puthoff, *Fundamentals of Quantum Electronics,* Wiley, New York (1969).
2. M. Asada and Y. Suematsu, *IEEE J. Quantum Electron.* **QE-21,** 434 (1985).
3. M. Asada, A. Kameyama, and Y. Suematsu, *IEEE J. Quantum Electron.* **QE-20,** 745 (1984).
4. Y. Arakawa and A. Yariv, *IEEE J. Quantum Electron.* **QE-22,** 1887 (1986).
5. C. Kittel, *Introduction to Solid State Phys., 4th ed.,* (1971).
6. L. I. Schitt, *Quantum Mechanics, 3rd ed.,* McGraw Hill, New York (1968).

ELEVEN

LASER GAIN AND SATURATION

The purpose of this chapter is to describe saturation phenomena in the inhomogeneously broadened gain spectrum of lasers. While giving some examples, we will use the laser equations to explain how the laser gain is saturated as the light intensity becomes strong. A phenomenon called "hole burning" will be mentioned, and in gas lasers, it will be related to the "Lamb dip" which was predicted by Willis Lamb. Furthermore, hole burning and saturation of gain in semiconductor lasers will be discussed. Readers who do not expect to become specialists in laser research and development need to understand the physical meaning of laser gain and saturation even if they do not delve into the detailed mathematics.

11.1. INHOMOGENEOUS BROADENING

Inhomogeneous broadening of lasers was mentioned in Chapter 10. There are two causes of inhomogeneous broadening: One is spectral inhomogeneity where the atoms and molecules or the electrons and hole pairs which provide laser gain do not have the same spectral characteristics, and the other is spatial inhomogeneity where a medium is inhomogeneously distributed in space.

As shown in Fig. 11.1, if the number of oscillators (generally atoms, molecules, and electrons are included) with angular frequencies in the range $d\Omega_i$ is expressed as dM, we find:

$$dM = N_v F(\Omega_i) d\Omega_i \tag{11.1}$$

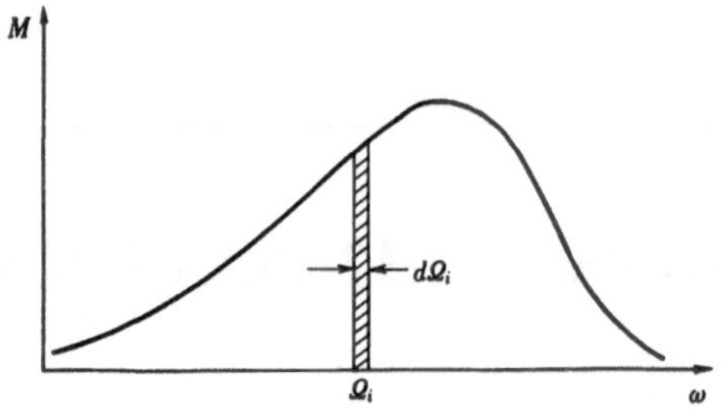

Figure 11.1. Distribution of molecules with inhomogeneous broadening.

where $F(\Omega_i)$ denotes the probability of the ith oscillator having an angular frequency, Ω_i.

If the total number of oscillators is written N_v, the following normalization is obtained:

$$N_v = \int dM \qquad \left(\because \int F(\Omega_i)d\Omega_i = 1\right) \tag{11.2}$$

11.2. HOLE BURNING

By assuming the angular frequency of the light and the resonance frequency of a molecule to be ω and Ω_i, respectively, the simultaneous equations for the diagonal elements of the density matrix for levels 1 and 2 become [from Eq. (9.38″)]:

$$\frac{\partial}{\partial t}(\rho_{22} - \rho_{11}) + \frac{1}{T_1}[(\rho_{22} - \rho_{11}) - \Delta\rho^e] = \frac{2}{\hbar\Omega_i}\langle\dot{\boldsymbol{\mu}}\rangle \cdot \boldsymbol{E}^{(l)} \tag{11.3}$$

Assuming a steady-state solution, i.e., $d/dt = 0$, we obtain

$$\Delta\rho - \Delta\rho^e = \frac{2T_1}{\hbar\Omega_i}\langle\dot{\boldsymbol{\mu}}\rangle \cdot \boldsymbol{E}^{(l)} \tag{11.4}$$

where

$$\Delta\rho = \rho_{22} - \rho_{11} \tag{11.4'}$$

If we write:

$$E^{(l)} = \frac{\mathbf{e}^{(l)}}{2} e^{j\omega t} + \text{c.c.} \tag{11.5}$$

$$\langle \mu \rangle = \frac{\langle \tilde{\mu} \rangle}{2} e^{j\omega t} + \text{c.c.} \tag{11.6}$$

then the right-hand side of Eq. (11.4) becomes:

$$\begin{aligned} \langle \dot{\mu} \rangle \cdot E^{(l)} &= \frac{1}{4}\frac{\partial}{\partial t}((\langle \tilde{\mu} \rangle e^{j\omega t} + \text{c.c.}) \cdot (e^{(l)} e^{j\omega t} + \text{c.c.}) \\ &\cong \frac{j\omega}{4}[\mathbf{e}^{(l)*} \cdot \langle \tilde{\mu} \rangle - \mathbf{e}^{(l)} \cdot \langle \tilde{\mu} \rangle^*] \end{aligned} \tag{11.7}$$

Here the rate of change in $\mathbf{e}^{(l)}$ and $\langle \mu \rangle$ has been assumed to be much slower than ω. Accordingly, Eq. (11.4) is written as:

$$\Delta\rho - \Delta\rho^e = \frac{j\omega T_1}{2\hbar\Omega_i}[\mathbf{e}^{(l)*} \cdot \langle \tilde{\mu} \rangle - \mathbf{e}^{(l)} \cdot \langle \tilde{\mu} \rangle^*] \tag{11.8}$$

The polarization equation is written from Eq. (9.38′) as:

$$\langle \ddot{\mu} \rangle + \frac{2}{T_2}\langle \dot{\mu} \rangle + \Omega_i^2 \langle \mu \rangle = -\frac{2\Omega_i}{\hbar} |\overline{\mu_{12}}|^2 \Delta\rho E^{(l)} \tag{11.9}$$

Using the same approximation, the left-hand side becomes:

$$\frac{1}{2}\left[(\Omega_i^2 - \omega^2) + \frac{2j\omega}{T_2}\right]\langle \tilde{\mu} \rangle e^{j\omega t} + \text{c.c.}$$

and assuming the light frequency is close to the resonance frequency:

$$\Omega_i^2 - \omega^2 = (\Omega_i - \omega)(\Omega_i + \omega) \cong 2\omega(\Omega_i - \omega) \tag{11.10}$$

Consequently, we obtain:

$$\omega\left[(\Omega_i-\omega)+\frac{j}{T_2}\right]\langle\tilde{\mu}\rangle e^{j\omega t}+\text{c.c.}$$

$$=-\frac{2\Omega_i}{\hbar}\,|\overline{\mu_{12}}|^2\Delta\rho\cdot\left(\frac{1}{2}\,\mathbf{e}^{(l)}e^{j\omega t}+\text{c.c.}\right)\tag{11.11}$$

then:

$$\left\{(\Omega_i-\omega)+\frac{j}{T_2}\right\}\langle\tilde{\mu}\rangle=-\frac{\Omega_i}{\hbar\omega}\cdot|\overline{\mu_{12}}|^2\Delta\rho\mathbf{e}^{(l)}$$

$$\therefore\langle\tilde{\mu}\rangle\cong-\frac{\Omega_i}{\hbar\omega}\cdot|\overline{\mu_{12}}|^2\Delta\rho\mathbf{e}^{(l)}\frac{1}{\Omega_i-\omega+j/T_2}\tag{11.12}$$

Substituting the above equation into Eq. (11.8) leads to:

$$\Delta\rho-\Delta\rho^e=-\frac{1}{\hbar^2}\,|\overline{\mu_{12}}|^2T_1T_2\Delta\rho\,|\mathbf{e}^{(l)}|^2\frac{1/T_2^2}{(\Omega_i-\omega)^2+1/T_2^2}$$

$$=-\frac{|\mathbf{e}^{(l)}|^2}{|\mathbf{e}_0^{(l)}|^2}\cdot\frac{1/T_2^2}{(\Omega_i-\omega)^2+1/T_2^2}\Delta\rho\tag{11.13}$$

where

$$|\mathbf{e}_0^{(l)}|^2\equiv\left[\frac{1}{\hbar^2}\,|\overline{\mu_{12}}|^2T_1T_2\right]-1\tag{11.14}$$

Rewriting Eq. (11.13) gives:

$$\Delta\rho=\Delta\rho^e\frac{1}{1+\dfrac{|\mathbf{e}^{(l)}|^2}{|\mathbf{e}_0^{(l)}|^2}\cdot\dfrac{1/T_2^2}{(\Omega_i-\omega)^2+1/T_2^2}}\tag{11.15}$$

Equation (11.15) proves that if optical power is so weak that we can write $|\mathbf{e}^{(l)}|^2\cong 0$, $\Delta\rho$ is equal to $\Delta\rho^e$. On the other hand, if the optical power is strong, $\Delta\rho\rightarrow 0$. This is a so-called gain saturation.

Now let us discuss the distribution probability of molecules again. The population differences $d\Delta N$ per unit angular frequency is written as:

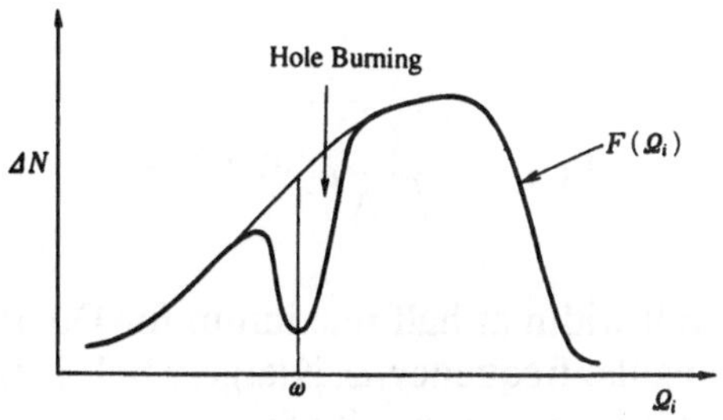

Figure 11.2. Hole burning.

$$\frac{d\Delta N}{d\Omega_i} = \Delta\rho\frac{dM}{d\Omega_i} = \Delta N^e \frac{F(\Omega_i)}{1 + \dfrac{|\mathbf{e}^{(l)}|^2}{|\mathbf{e}_0^{(l)}|^2}\cdot\dfrac{1/T_2^2}{(\Omega_i - \omega)^2 + 1/T_2^2}} \tag{11.16}$$

where we have defined the equilibrium population difference $\Delta N^e = \Delta\rho^e \cdot N_v$.

When the optical power becomes strong at a particular frequency, a hole is made in the gain distribution expressed by $F(\Omega_i)$. This is called hole burning, and the depth of the hole depends on the optical power (Fig. 11.2).

Next, let us discuss an example of hole burning in a gas laser. The apparent resonant frequencies of molecules in a gas are shifted by the Doppler effect as a result of their thermal motion. That is, if the velocity component in the same direction as that of a traveling electromagnetic wave is v, then the resonant frequency observed in the laboratory frame is:

$$\Omega_i = \omega_0\left(1 + \frac{v}{c}\right) \tag{11.17}$$

where c denotes the speed of light.

The velocity distribution for thermal equilibrium is the Maxwell–Boltzmann distribution, and the spectral distribution can be derived from this by using Eq. (11.17) to give:

$$dM = N_v \frac{1}{\sqrt{\pi}\Delta} e^{-(\Omega_i - \omega_0)^2/\Delta^2} d\Omega_i = N_v F(\Omega_i) d\Omega_i \tag{11.18}$$

This frequency broadening is called Doppler broadening. The linewidth $\Delta\nu_D$ or half-width at half maximum of the Doppler broadening is written as:

$$\Delta\nu_D = \frac{1}{2\pi} \times \sqrt{\ln 2}\,\Delta \tag{11.19}$$

here we have:

$$F(\Omega_i) = \frac{1}{\sqrt{\pi}\,\Delta} e^{-(\Omega_i - \omega_0)^2/\Delta^2} \tag{11.18'}$$

Sometimes we use full width at half maximum for Doppler broadening. If the intensity at the angular frequency, ω, is large, a hole in the Doppler profile is made around ω. This hole is the result of saturation of the Doppler-shifted group of molecules which occurs when only these molecules are resonant with the propagating light, as is shown in Fig. 11.3b. In a resonator, if $\omega \neq \omega_0$, a second hole is made near ω' symmetrical to ω_0 because of the backward propagating beam. Then we can write:

$$\frac{d\Delta N}{d\Omega_i} = \Delta N^e \frac{\dfrac{1}{\sqrt{\pi}\,\Delta} e^{-(\Omega_i - \omega_0)^2/\Delta^2}}{1 + \dfrac{|\mathbf{e}^{(l)}|^2}{|\mathbf{e}_0^{(l)}|^2}\left[\dfrac{1/T_2^2}{(\Omega_i - \omega)^2 + 1/T_2^2} + \dfrac{1/T_2^2}{(\Omega_i - \omega')^2 + 1/T_2^2}\right]} \tag{11.20}$$

where $\omega' = 2\omega_0 - \omega$.

If the oscillation angular frequency, ω, of a laser approaches ω_0, the two holes are superimposed, thus causing a larger effect. By reflecting a laser beam back through a Doppler-broadened medium, this central hole can be probed in absorption. Lamb carried out a calculation using a perturbation method to theoretically predict this effect.[1] The hole in the absorption profile is therefore called the Lamb dip in his honor. This effect was later confirmed experimentally[2] and is now commonly used for laser stabilization.

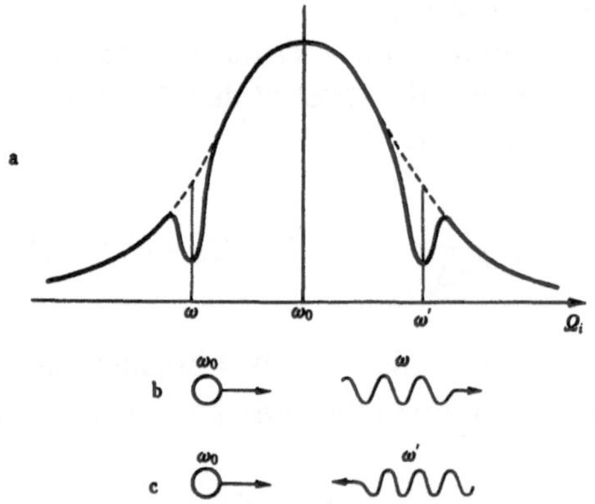

Figure 11.3. Two holes in a system with inhomogeneous broadening.

11.3. SATURATION OF LIGHT OUTPUT

Although the hole in Eq. (11.20) is made, it cannot easily be observed from outside of the resonator cavity. One possible way we can observe it is to inject light with different frequencies into an oscillating laser and observe the difference in the gain factor at these frequencies, but that is not an easy experiment. However, the hole in the gain can be directly recognized through the change in the light output (saturation characteristic) which occurs when the frequency is tuned by sweeping the laser resonance frequency, or through the competition between oscillating modes. Observations such as these allow us to determine the characteristics of a laser.

Here we discuss the light output of a laser which has an inhomogeneous gain profile. The electric field of the light is written from Eq. (9.38) as:

$$\ddot{\mathbf{E}} + \frac{1}{\tau p}\dot{\mathbf{E}} + \omega_c^2\mathbf{E} = -\frac{1}{\varepsilon}\ddot{\mathbf{P}} \tag{11.21}$$

In the same manner as in Eq. (10.7), we assume a single-frequency component, so the macroscopic polarization may be expressed as:

$$\mathbf{P} = \tfrac{1}{2}\mathbf{p}\exp^{(j\omega t)} + \text{c.c.} \tag{11.22}$$

and the electric field is written:

$$\mathbf{E} = \tfrac{1}{2}\mathbf{e}\exp^{(j\omega t)} + \text{c.c.} \tag{11.23}$$

By substituting Eqs. (11.22) and (11.23) into Eq. (11.21), we obtain:

$$[(\omega_c^2 - \omega^2) + j\omega/\tau_p]\mathbf{e} \cong +\frac{\omega^2}{\varepsilon}\mathbf{p} \tag{11.24}$$

where the temporal change of **p** and **e** has been assumed to be much slower than ω. If **e** is a real number, from the imaginary part of Eq. (11.24), we obtain:

$$\frac{\mathbf{e}}{\tau_p} \cong \frac{\omega}{\varepsilon}\text{Im}[\mathbf{p}] \tag{11.25}$$

This is the condition for steady-state laser oscillation.

We also obtain the imaginary part of $\langle\mu\rangle$ from Eq. (11.12) as:

$$\mathrm{Im}[\langle\tilde{\mu}\rangle] = +\frac{\Omega_i}{\hbar\omega}|\overline{\mu_{12}}|^2\Delta\rho\mathbf{e}^{(l)}\frac{1/T_2}{(\Omega_i-\omega)^2+1/T_2^2} \tag{11.26}$$

Then, by using Eq. (11.18), we can write:

$$\begin{aligned}\mathrm{Im}[\mathbf{p}] &= \int \mathrm{Im}\,[\langle\tilde{\mu}\rangle]dM = N_v\int \mathrm{Im}[\langle\tilde{\mu}\rangle]F(\Omega_i)d\Omega_i \\ &= \frac{N_v}{\hbar\omega}|\overline{\mu_{12}}|^2\int \Delta\rho\mathbf{e}^{(l)}\frac{\Omega_i/T_2}{(\Omega_i-\omega)^2+1/T_2^2}F(\Omega_i)d\Omega_i \end{aligned} \tag{11.27}$$

If the saturation is not very strong, and the change in $\Delta\rho$ and $\mathbf{e}^{(l)}$ compared with the optical frequency is relatively slow, then we can write:

$$\mathrm{Im}[\mathbf{p}] \cong \frac{N_v}{\hbar\omega}|\overline{\mu_{12}}|^2\Delta\rho\mathbf{e}^{(l)}\int\frac{\Omega_i/T_2}{(\Omega_i-\omega)^2+1/T_2^2}F(\Omega_i)d\Omega_i \tag{11.28}$$

Furthermore, if the homogeneous width, $1/T_2$, is assumed to be narrower than the inhomogeneous width, Δ, the following approximation is reasonably obtained:

$$\frac{T_2}{T_2^2(\Omega_i-\omega)^2+1} \cong \pi\delta(\Omega_i-\omega) \qquad (T_2\Delta \gg 1) \tag{11.29}$$

where δ denotes a Dirac delta function whose normalization condition is:

$$\int \delta(\Omega-\omega)d\Omega = 1$$

Then we can write:

$$\begin{aligned}\mathrm{Im}[\mathbf{p}] &\cong \frac{N_v}{\hbar\omega}|\overline{\mu_{12}}|^2\Delta\rho\mathbf{e}^{(l)}\int \pi\delta(\Omega_i-\omega)\Omega_i F(\Omega_i)d\Omega_i \\ &= \frac{N_v\pi}{\hbar}|\overline{\mu_{12}}|^2\Delta\rho\mathbf{e}^{(l)}F(\omega)\end{aligned} \tag{11.30}$$

In order to obtain an oscillation condition, we substitute Eq. (11.30) into Eq. (11.5) and use $\mathbf{e}^{(l)} = \sqrt{\mathcal{L}}e$, which gives:

$$\frac{1}{\tau_p} = \frac{\pi\omega}{\hbar\varepsilon} \cdot |\overline{\mu_{12}}|^2 \sqrt{\mathcal{L}} N_v \Delta\rho_{\text{th}} F(\omega) \tag{11.31}$$

where $\mathcal{L}$ is the Lorentz local field correction factor.[3]

The threshold electric power W_{th} which is required for pumping is written:

$$W_{\text{th}} = \tfrac{1}{2} \hbar\omega / T_1 \cdot \Delta N_{\text{th}}^e \tag{11.32}$$

Therefore, if we express $N_v \Delta\rho_{\text{th}}^e$ as ΔN_{th}^e in Eq. (11.31), the following relationship is obtained:

$$W_{\text{th}} = \frac{1}{2} \frac{\hbar^2 \varepsilon}{\pi T_1 \tau_p |\overline{\mu_{12}}|^2 \sqrt{\mathcal{L}}} \cdot \frac{1}{F(\omega)} \tag{11.33}$$

By tuning the frequency to $\omega = \omega_0$, $F(\omega_0)$ becomes equal to $1/(\sqrt{\pi}\Delta)$, and the minimum value for the excitation energy is obtained:

$$W_{\text{th}}(\omega_0) = \frac{1}{2} \cdot \frac{\hbar^2 \varepsilon}{\pi T_1 \tau_p |\overline{\mu_{12}}|^2 \sqrt{\mathcal{L}}} \cdot \sqrt{\pi}\Delta \tag{11.33$'$}$$

In an operating laser the output power is limited by saturation. If we put:

$$|\mathbf{e}^{(l)}|^2 / |\mathfrak{e}_0^{(l)}|^2 \equiv U$$

and deal with two holes, as shown in Eq. (11.20), we can write a new expression by using Eqs. (11.20), (11.25), and (11.31):

$$1 = \frac{1}{\omega\pi} \frac{W}{W_{\text{th}}(\omega_0)} \int \frac{\Omega_i / T_2}{(\Omega_i - \omega)^2 + 1/T_2^2}$$

$$\times \frac{\sqrt{\pi}\Delta F(\Omega_i) d\Omega_i}{1 + U\left[\dfrac{1/T_2^2}{(\Omega_i - \omega)^2 + 1/T_2^2} + \dfrac{1/T_2^2}{(\Omega_i - \omega')^2 + 1/T_2^2}\right]} \tag{11.34}$$

where

$$W = \frac{1}{2} N_v \Delta\rho^e \frac{\hbar\omega}{T_1}$$

If we use the approximation $\omega + \omega' = 2\omega_0$, we find:

$$\frac{W}{W_{\text{th}}(\omega_0)} = \left[\frac{1}{\omega\pi} \int \frac{\sqrt{\pi}\Delta T_2 \Omega_i F(\Omega_i) d\Omega_i}{T_2^2(\Omega_i - \omega)^2 + 1 + U\left\{1 + \dfrac{T_2^2(\Omega_i - \omega)^2 + 1}{T_2^2(\Omega_i - 2\omega_0 + \omega)^2 + 1}\right\}} \right]^{-1} \quad (11.34')$$

In the denominator of this integrand, the term including U is assumed to be smaller than the first term and is abbreviated h. If $\Delta \cdot T_2 \gg 1$, then

$$\frac{T_2}{T_2^2(\Omega_i - \omega)^2 + 1 + h} = \frac{T_2/\sqrt{1+h}}{\left(\dfrac{\Omega_i T_2}{\sqrt{1+h}}\right)^2 \left(1 - \dfrac{\omega}{\Omega_i}\right)^2 + 1} \cdot \frac{1}{\sqrt{1+h}}$$

$$\cong \pi\delta(\Omega_i - \omega) \frac{1}{\sqrt{1+h}} \quad (11.34'')$$

where $\delta(\Omega_i - \omega)$ is again the Dirac delta function. Consequently, the integral becomes:

$$\frac{W}{W_{\text{th}}(\omega_0)} \cong \left\{ \sqrt{\pi}\Delta F(\omega) \frac{1}{\sqrt{1 + h(\Omega_i - \omega)}} \right\}^{-1} \quad (11.35)$$

By approximating,

$$\sqrt{1+h} \cong 1 + \tfrac{1}{2}h$$

we can write

$$\frac{W}{W_{\text{th}}(\omega_0)} \cdot \sqrt{\pi}\Delta F(\omega) = 1 + \frac{1}{2} U \left[1 + \frac{1}{4T_2^2(\omega_0 - \omega)^2 + 1} \right] \quad (11.36)$$

from which we find:

$$U \cong \frac{2}{1+\dfrac{1}{4T_2^2(\omega_0-\omega)^2+1}}\cdot\left\{\frac{W}{W_{\rm th}(\omega_0)}\sqrt{\pi}\,\Delta F(\omega)-1\right\} \qquad (11.37)$$

It is seen from the above equations that the relative optical power, U, is distributed around the center of $F(\omega)$ when the excitation is weak, but hole burning occurs at the center of the line when the excitation is strong, as is shown in Fig. 11.4. By substituting for $F(\omega)$ from Eq. (11.18′), we get:

$$U = \frac{2}{1+\dfrac{1}{1+4T_2^2(\omega-\omega_0)^2}}\left\{\frac{W}{W_{\rm th}(\omega_0)}e^{-(\omega-\omega_0)^2/\Delta^2}-1\right\} \qquad (11.37')$$

If $\omega=\omega_0$, U becomes:

$$U = \frac{W}{W_{\rm th}(\omega_0)}-1 \qquad (11.37'')$$

From Eq. (11.37′), we see that the Doppler half-width, $\Delta\omega_0$ (the half-width at half maximum), is equal to:

$$\Delta\omega_0 = \sqrt{\ln 2}\,\Delta = \sqrt{0.69}\,\Delta$$

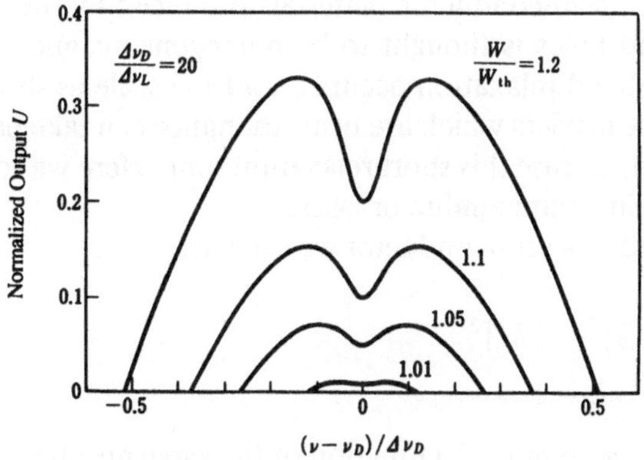

Figure 11.4. Lamb dip in a gas laser.

From the expression, $4T_2(\omega - \omega_0) = 1$, the Lorentzian half-width of the Lamb dip, $\Delta\omega_0$ is equal to $0.5/T_2$. Since we have assumed:

$$\Delta \gg 1/T_2$$

the Lorentz dip with the width $0.5/T_2$ is formed at the center of a Gaussian function with a half-width at half maximum of 0.83Δ.

Rewriting Eq. (11.37′) using $\omega = 2\pi\nu$, $\omega_0 = 2\pi\nu_0$, $\Delta\omega_D = 2\pi\Delta\omega_D$, and $\Delta\omega_L = 2\pi\Delta\nu_L$ we find:

$$U = \frac{2}{1 + \dfrac{1}{1 + (\nu - \nu_0)^2/\Delta\nu_L^2}} \cdot \left[\frac{W}{W_{th}(\nu_0)} e^{-\ln 2(\nu - \nu_0)^2/\Delta\nu_D^2} - 1 \right] \qquad (11.37''')$$

Here, if $x = (\nu - \nu_0)/\Delta\nu_D$, we can write:

$$U = \frac{2}{1 + \dfrac{1}{1 + x^2(\Delta\nu_D/\Delta\nu_L)^2}} \cdot \left[\frac{W}{W_{th}(\nu_0)} e^{-\ln 2 \cdot x^2} - 1 \right] \qquad (11.37'''')$$

Figure 11.4 shows U for the case where $\Delta\nu_D/\Delta\nu_L = 20$. Readers should understand that the dip becomes increasingly noticeable as the relative pump power, $W/W_{th}(\nu_0)$, becomes large.

11.4. GAIN AND SATURATION IN SEMICONDUCTOR LASERS

The gain in semiconductor materials produced by the recombination of electrons and holes is thought to be homogeneous and relatively wide. However, intraband relaxation occurs on a time scale as short as 10^{-13} sec and many of the carriers which are near resonance can take part in the stimulated emission because this short relaxation time. Here we consider the saturation of gain in a semiconductor laser.

For the case of a semiconductor we can write:

$$|\overline{\mu_{12}}|^2 = |M|^2$$

$$L = 1$$

$$\omega_0 = \omega_{21} \qquad \text{(function of the wave number } \mathbf{K}\text{)} \qquad (11.38)$$

Then Eq. (9.38′) becomes:

$$\langle \ddot{\mu} \rangle + \frac{2}{T_2}\langle \dot{\mu} \rangle + \omega_{21}^2 \langle \mu \rangle = -B(\rho_{22} - \rho_{11})\mathbf{E} \tag{11.39}$$

Here,

$$B \equiv \frac{2\omega_{21}}{\hbar} |\overline{\mu_{12}}|^2 = \frac{2\omega_{21}}{\hbar} |M|^2 \tag{11.40}$$

From Eqs. (9.38″) and (9.38) the following equations are obtained:

$$\frac{d}{dt}(\rho_{22} - \rho_{11}) + \frac{1}{T_1}[(\rho_{22} - \rho_{11}) - \Delta\rho^e] = \frac{2}{\hbar\omega_{21}}\langle \dot{\mu} \rangle \cdot \mathbf{E} \tag{11.41}$$

$$\ddot{\mathbf{E}} + \frac{1}{\tau_p}\dot{\mathbf{E}} + \omega_c^2 \mathbf{E} = -\frac{1}{\varepsilon}\ddot{\mathbf{P}} \tag{11.42}$$

In the case of semiconductor lasers, we have to consider the so-called intraband relaxation. The electrons in the conduction band relax to adjacent levels having different values of k and different energy with a relaxation time which can be as short as 10^{-13} sec. The same thing occurs for holes in the valence band. This relaxation is the result of carrier–carrier interactions, phonon scattering, and other relaxation mechanisms.

This intraband relaxation can be described by the following equations:

$$\frac{\partial \rho_{22}}{\partial t} = -\frac{\rho_{22} - \rho'_{22}}{T_c} \tag{11.43}$$

$$\frac{\partial \rho_{11}}{\partial t} = -\frac{\rho_{11} - \rho'_{11}}{T_v} \tag{11.44}$$

$$\frac{\partial \rho_{12}}{\partial t} = -\frac{1}{T_2}\rho_{12} \tag{11.45}$$

where

$$\frac{1}{T_2} = \frac{1}{2}\left(\frac{1}{T_c} + \frac{1}{T_r}\right) \tag{11.46}$$

and ρ'_{ii} expresses the density of states in quasi-Fermi thermal equilibrium.[4,5] We therefore have to modify Eq. (11.41) to read:

$$\frac{d}{dt}\Delta\rho + \frac{1}{T_1}(\Delta\rho - \Delta\rho^e) + \frac{1}{T_2}(\Delta\rho - \Delta\rho') = \frac{2}{\hbar\omega_{12}}\langle\dot{\mu}\rangle\cdot\mathbf{E} \quad (11.41')$$

where

$$\Delta\rho' = \rho'_{22} - \rho'_{11}$$

is the state difference in the quasi-thermal equilibrium. The distributions of carriers in equilibrium are shown in Fig. 11.5.

The macroscopic polarization is written:

$$\mathbf{P} = \int d^3\mathbf{k}\langle\mu\rangle \quad (11.47)$$

As before, we can write:

$$\mathbf{E} = \frac{\mathbf{e}}{2}e^{j\omega t} + \frac{\mathbf{e}^*}{2}e^{-j\omega t} \quad (11.48)$$

$$\langle\mu\rangle = \frac{\langle\tilde{\mu}\rangle}{2}e^{j\omega t} + \frac{\langle\tilde{\mu}\rangle^*}{2}e^{-j\omega t} \quad (11.49)$$

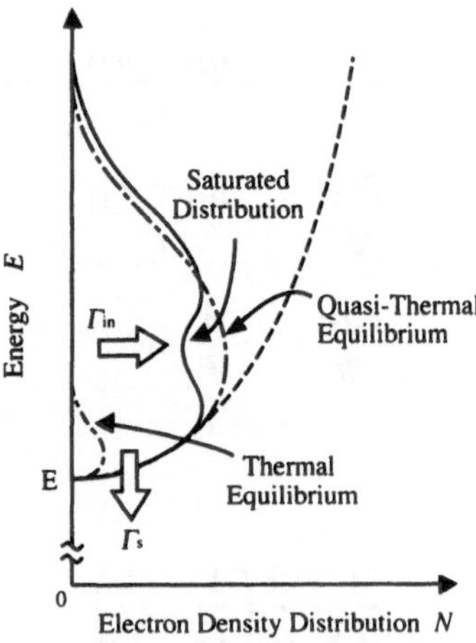

Figure 11.5. Quasi-Fermi distribution of carriers.

For this case we deal with a single-mode oscillation:

$$\frac{d}{dt}\langle\mu\rangle\cdot\mathbf{E} = \frac{j\omega}{4}[\langle\tilde{\mu}\rangle\cdot\mathbf{e}^* - \langle\tilde{\mu}\rangle^*\cdot\mathbf{e}] \qquad (11.50)$$

Consequently, for a steady-state condition, we obtain:

$$\frac{1}{T_2}[(\rho_{22}-\rho_{11}) - \Delta\rho'] = \frac{2}{\hbar\omega_{21}}\cdot\frac{j\omega}{4}[\langle\tilde{\mu}\rangle\cdot\mathbf{e}^* - \langle\tilde{\mu}\rangle^*\cdot\mathbf{e}] \qquad (11.51)$$

where we have assumed that $T_1 > T_2$ and the intraband relaxation term dominates. Returning to Eq. (11.39) and making the approximation:

$$\langle\ddot{\mu}\rangle \cong (j\omega)^2\langle\mu\rangle \;\therefore \frac{1}{2}\langle\tilde{\mu}\rangle\left[-\omega^2 + \frac{2}{T_2}j\omega + \omega_{21}^2\right]e^{j\omega t} + \text{c.c.}$$

$$= -B(\rho_{22}-\rho_{11})\cdot\frac{1}{2}\mathbf{e}e^{j\omega t} + \text{c.c.}$$

we find:

$$\therefore \langle\tilde{\mu}\rangle = \frac{1}{\omega_{21}^2 - \omega^2 + (2/T_2)j\omega}(-B)(\rho_{22}-\rho_{11})\mathbf{e}$$

$$\cong \frac{1}{j(\omega-\omega_{21}) + 1/T_2}\frac{j}{\hbar}|M|^2(\omega_{21}/\omega)(\rho_{22}-\rho_{11})\mathbf{e} \qquad (11.52)$$

This expression can be substituted into Eq. (11.51) to yield:

$$\frac{1}{T_2}[(\rho_{22}-\rho_{11}) - \Delta\rho']$$

$$= \frac{j\omega}{2\hbar\omega_{21}}\cdot\left[\frac{1}{\omega_{21}^2 - \omega^2 + (2/T_2)j\omega} - \frac{1}{\omega_{21}^2 - \omega^2 + (2/T_2)(-j\omega)}\right]$$

$$\times(-B)(\rho_{22}-\rho_{11})|\mathbf{e}|^2 = j\omega\cdot\frac{|M|^2}{\hbar^2}\left[\frac{4j\omega/T_2}{(\omega_{21}^2-\omega^2)^2 + 4\omega^2/T_2^2}\right]$$

$$\times(\rho_{22}-\rho_{11})|\mathbf{e}|^2 = -\frac{4|M|^2\omega^2}{\hbar^2}\frac{1/T_2}{(\omega_{21}^2-\omega^2)^2 + 4\omega^2/T_2^2}$$

$$\times(\rho_{22}-\rho_{11})|\mathbf{e}|^2 \qquad (11.53)$$

This equation can be rearranged to write:

$$\rho_{22} - \rho_{11} - \Delta\rho^e = -\frac{4|M|^2 T_1}{\hbar^2} \times \frac{\omega^2(1/T_2)}{(\omega_{21}^2 - \omega^2)^2 + 4\omega^2/T_2^2}(\rho_{22} - \rho_{11})|\mathbf{e}|^2$$

$$\cong -\frac{|\mathbf{e}|^2}{|\mathbf{e}_s|^2}(\rho_{22} - \rho_{11})\frac{1/T_2}{(\omega_{21} - \omega)^2 + 1/T_2^2} \tag{11.54}$$

where we have used the approximation:

$$\frac{1}{(\omega_{21}^2 - \omega^2)^2 + 4\omega^2/T_2^2} \cong \frac{1}{[(\omega_{21} - \omega)^2 + 1/T_2^2](4\omega^2)} \tag{11.55}$$

and

$$|\mathbf{e}_s|^2 = \frac{\hbar^2}{|M|^2 T_2} \tag{11.56}$$

With one further iteration, the following equation is obtained:

$$(\rho_{22} - \rho_{11}) = \frac{\Delta\rho'}{1 + \dfrac{|\mathbf{e}|^2}{|\mathbf{e}_s|^2}\dfrac{1/T_2^2}{(\omega_{21} - \omega)^2 + 1/T_2^2}} \tag{11.57}$$

Using similar approximations, the microscopic polarization becomes:

$$\langle\tilde{\mu}\rangle = \frac{1}{\omega_{21}^2 - \omega^2 + 2j\omega/T_2}(-B)\frac{\mathbf{e}\Delta\rho'}{1 + \dfrac{|\mathbf{e}|^2}{|\mathbf{e}_s|^2}\dfrac{1/T_2^2}{(\omega_{21} - \omega)^2 + 1/T_2^2}} \tag{11.58}$$

Note that:

$$\frac{1}{\omega_{21}^2-\omega^2+2j\omega/T_2}\cong\frac{1}{[\omega_{21}-\omega+j/T_2](2\omega)}$$
$$=\frac{\omega_{21}-\omega-j/T_2}{(\omega_{21}-\omega)^2+1/T_2^2}\cdot\frac{1}{2\omega} \qquad (11.58')$$

and therefore we can write:

$$\langle\tilde{\mu}\rangle=-\frac{2\omega_{21}}{\hbar}|M|^2\frac{1}{2\omega}\cdot\frac{\omega_{21}-\omega-j/T_2}{(\omega_{21}-\omega)^2+1/T_2^2}$$
$$\times\frac{\mathbf{e}\Delta\rho'}{1+\dfrac{|\mathbf{e}|^2}{|\mathbf{e}_s|^2}\dfrac{1/T_2^2}{(\omega_{21}-\omega)^2+1/T_2^2}} \qquad (11.59)$$

$$\mathrm{Re}\{\langle\tilde{\mu}\rangle\}=-\frac{\omega_{21}}{\hbar\omega}|M|^2\frac{\omega_{21}-\omega}{(\omega_{21}-\omega)^2+1/T_2^2}$$
$$\times\frac{\mathbf{e}\Delta\rho'}{1+\dfrac{|\mathbf{e}|^2}{|\mathbf{e}_s|^2}\dfrac{1/T_2^2}{(\omega_{21}-\omega)^2+1/T_2^2}} \qquad (11.60)$$

$$\mathrm{Im}\{\langle\tilde{\mu}\rangle\}=\frac{\omega_{21}}{\hbar\omega}|M|^2\frac{1/T_2}{(\omega_{21}-\omega)^2+1/T_2^2}$$
$$\times\frac{\mathbf{e}\Delta\rho'}{1+\dfrac{|\mathbf{e}|^2}{|\mathbf{e}_s|^2}\dfrac{1/T_2^2}{(\omega_{21}-\omega)^2+1/T_2^2}} \qquad (11.61)$$

These expressions can be put in terms of the macroscopic polarization, **P**, using Eq. (11.47). It is convenient to separate these expressions into real and imaginary parts:

$$\mathrm{Re}\{\mathbf{P}\}=\varepsilon_0\chi'\mathbf{e} \qquad (11.62)$$
$$\mathrm{Im}\{\mathbf{P}\}=\varepsilon_0\chi''\mathbf{e} \qquad (11.63)$$

which lead to the following relations:

$$\chi' = \int d^3\mathbf{k}(-)\frac{\omega_{21}}{\hbar\omega^{\varepsilon_0}_{0}}\,|M|^2\frac{\omega_{21}-\omega}{(\omega_{21}-\omega)^2+1/T_2^2}$$

$$\times\frac{\bar{N}\Delta\rho'}{1+\dfrac{|\mathbf{e}|^2}{|\mathbf{e}_s|^2}\dfrac{1/T_2^2}{(\omega_{21}-\omega)^2+1/T_2^2}} \tag{11.64}$$

$$\chi'' = \int d^3\mathbf{k}\,\frac{\omega_{21}}{\hbar\omega\varepsilon_0}\,|M|^2\frac{1/T_2}{(\omega_{21}-\omega)^2+1/T_2^2}$$

$$\times\frac{\bar{N}\Delta\rho'}{1+\dfrac{|\mathbf{e}|^2}{|\mathbf{e}_s|^2}\dfrac{1/T_2^2}{(\omega_{21}-\omega)^2+1/T_2^2}} \tag{11.65}$$

where N denotes the carrier density. The relationship between the real and imaginary parts is similar to the well-known Kramers–Kronig relation.

From the equation for the electric field of light, we can write:

$$\dot{\mathbf{E}} = \left(\frac{\dot{\mathbf{e}}}{2}+j\omega\frac{\mathbf{e}}{2}\right)\exp(j\omega t)+\text{c.c.} \tag{11.66}$$

which leads to:

$$\ddot{\mathbf{E}} = \left(\frac{\ddot{\mathbf{e}}}{2}+j\omega\frac{\dot{\mathbf{e}}}{2}+j\omega\frac{\dot{\mathbf{e}}}{2}+(j\omega)^2\frac{\mathbf{e}}{2}\right)\exp(j\omega t)+\text{c.c.} \tag{11.67}$$

Returning to Eq. (11.42), we find:

$$\left[\left(\frac{\ddot{\mathbf{e}}}{2}+j\omega\dot{\mathbf{e}}-\omega^2\frac{\mathbf{e}}{2}\right)+\left(\frac{\dot{\mathbf{e}}}{2}\cdot\frac{1}{\tau_p}+j\frac{\omega}{\tau_p}\frac{\mathbf{e}}{2}\right)+\omega_c^2\frac{\mathbf{e}}{2}\right]\exp(j\omega t)+\text{c.c.}$$

$$\cong\left\{\left(j\omega+\frac{1}{2\tau_p}\right)\dot{\mathbf{e}}+\left[\omega_c^2-\omega^2+j\frac{\omega}{\tau_p}\right]\frac{\mathbf{e}}{2}\right\}\exp(j\omega t)+\text{c.c.}$$

$$\triangleq j\omega\left\{\dot{\mathbf{e}}+\left(\frac{\omega_c^2-\omega^2}{j\omega}+\frac{1}{\tau_p}\right)\frac{\mathbf{e}}{2}\right\}\exp(j\omega t)+\text{c.c.}$$

$$=-\frac{1}{\varepsilon_0}\bar{N}\left\{\frac{\langle\dot{\tilde{\mu}}\rangle}{2}j\omega-\frac{\langle\tilde{\mu}\rangle}{2}\omega^2\right\}\exp(j\omega t)+\text{c.c.}\quad(11.68)$$

Consequently,

$$\frac{d\mathbf{e}}{dt}+\frac{1}{2}[(\omega_c^2-\omega^2)/j\omega+1/\tau_p]\mathbf{e}=\frac{\omega}{2j\varepsilon_0}\int d^3\mathbf{k}\bar{N}\langle\tilde{\mu}\rangle$$

$$=\frac{\omega}{j2n_r^2}(\chi'+j\chi'')\mathbf{e}\qquad(11.69)$$

and therefore,

$$\therefore\frac{d\mathbf{e}}{dt}+\frac{1}{2}\left\{\left(\frac{1}{\tau_p}-\tilde{G}\right)+\left[(\omega_c^2-\omega^2)/j\omega-\frac{\omega}{jn_r^2\chi'}\right]\right\}\mathbf{e}=0\qquad(11.70)$$

Accordingly, the nonlinear gain, G, can be calculated from the following equation:

$$\tilde{G}=\frac{\omega}{n_r^2}\chi''\qquad(11.71)$$

where n_r denotes the refractive index of the medium. Substituting Eq. (11.65) leads to an expression of the gain factor per unit time:

$$\tilde{G}=\frac{\omega}{n_r^2}\chi''=\frac{\omega}{\varepsilon_0 n_r^2}\int d^3\mathbf{k}\frac{\omega_{21}}{\hbar\omega}|M|^2\times\frac{1/T_2}{(\omega_{21}-\omega)^2+1/T_2^2}$$

$$\times\frac{\bar{N}\Delta\rho'}{1+\dfrac{|\mathbf{e}|^2}{|\mathbf{e}_s|^2}\dfrac{1/T_2^2}{(\omega_{21}-\omega)^2+1/T_2^2}}\qquad(11.72)$$

In semiconductors, the energy of the electrons in the conduction bands is expressed by:

$$E_c = E_{10} + \frac{\hbar^2 k_c^2}{2m_c^*} = E_{10} + \frac{m_v^*}{m_c^* + m_v^*}(E_{21} - E_g) \qquad (11.73)$$

whereas the energy for holes is given by:

$$E_v = E_{v0} - \frac{\hbar^2 k_v^2}{2m_v^*} = E_{v0} - \frac{m_c^*}{m_c^* + m_v^*}(E_{21} - E_g) \qquad (11.73')$$

The energy difference, $E_{21} = \hbar\omega_{21}$, is:

$$E_{21} = E_g + E_c + E_v = \hbar\omega_{21} \qquad (11.74)$$

where E_g is the band gap energy.

By using the parabolic density of states in a semiconductor, the integral over $\mathbf{k}$ may be expressed as an integral over E:

$$\frac{2}{(2\pi)^3}\int d^3\mathbf{k} \times \int dE_{21}\bar{g}(E_{21}) \times \qquad (11.75)$$

where

$$\bar{g}(E_{21})dE_{21} = \left(\frac{2m_c^* m_v^*}{m_c^* + m_v^*}\right)^{3/2} \frac{1}{2\pi^2\hbar^3}\sqrt{E_{21} - E_g}\,dE_{21} \qquad (11.76)$$

Because $\Delta\rho' N$ is considered to be in thermal equilibrium, we can write:

$$\Delta\rho'\bar{N} = [f_c(E_{21}) - f_v(E_{21})] \times \frac{2}{(2\pi)^3} \qquad (11.77)$$

Equations (11.75) and (11.76) can be used to rewrite Eq. (11.72) to give:

$$\tilde{G} = \frac{\omega}{n_r^2}\chi'' = \frac{\omega}{\varepsilon_0 n_r^2}\int_{E_g}^{\infty} dE_{21} M^2 \bar{g}(E_{21})[f_c - f_v](\hbar/T_2)$$

$$\times\left[1+\frac{|e|^2}{|e_s|^2}\frac{(\hbar/T_2)^2}{(\hbar\omega-E_{21})^2+(\hbar/T_2)^2}\right]^{-1}$$

$$\times\frac{1}{(\hbar\omega-E_{21})^2+(\hbar/T_2)^2} \quad (11.78)$$

When we expand this equation in terms of $|e|^2/|e_s|^2$, we have:

$$\tilde{G}=A-B|e|^2/|e_s|^2+C|e|^4/|e_s|^4+\cdots \quad (11.79)$$

The linear portion of this gain is obtained from the expression:

$$A=\frac{\omega}{\varepsilon_0 n_r^2}\int_{E_g}^{\infty}\frac{M^2\bar{g}(f_c-f_v)(\hbar/T_2)}{(\hbar\omega-E_{21})^2+(\hbar/T_2)^2}dE_{21} \quad (11.80)$$

The saturation coefficient B is given by:

$$B=\frac{2\hbar\omega^2}{\varepsilon_0^2 n_r^4}\int_{E_g}^{\infty}\frac{M^4\bar{g}(f_c-f_v)(\hbar/T_2)}{[(\hbar\omega-E_{21})^2+(\hbar/T_2)^2]^2}dE_{21} \quad (11.81)$$

If we again approximate the Lorentzian function as:

$$\frac{\hbar/T_2}{(\hbar\omega-E_{21})^2+(\hbar/T_2)^2}=\delta(E_{21}-\hbar\omega) \quad (11.82)$$

then we can express the coefficients A and B as:

$$A\cong\bar{g}(f_c-f_v)|_{\hbar\omega=E_{21}}\cong a(N-N_g) \quad (11.83)$$

$$B\cong\frac{2\hbar\omega}{\varepsilon_0 n_r^2}\left(\frac{T_2}{\hbar}\right)^2 M^2 a(N-N_s) \quad (11.84)$$

where N_g and N_s are a transparency carrier density and a saturation carrier density, respectively. The gain, g, per unit length can be found by letting $g\Delta z = G\Delta t$, which gives:

$$g=\frac{n_r}{c}\bar{G}=\sqrt{\frac{\mu_0}{\varepsilon_0}}\frac{1}{\hbar n_r}\int d^3\mathbf{k}\frac{\omega_{21}/T_2}{(\omega_{21}-\omega)^2+1/T_2^2}|M|^2\bar{N}\Delta\rho^e \quad (11.85)$$

Accordingly, the linear gain coefficient is:

$$g = \sqrt{\frac{\mu_0}{\varepsilon_0}} \frac{1}{\hbar n_r} \left(\frac{2m_c^* m_v^*}{m_c^* + m_v^*} \right)^{3/2} \frac{1}{2\pi^2 \hbar^3} \times \int_{E_g}^{\infty} dE_{21} \sqrt{E_{21} - E_g}$$

$$\times [f_c(E_{21}) - f_v(E_{21})] \frac{E_{21}(\hbar / T_2) |M|^2}{(E_{21} - \hbar\omega)^2 + \hbar^2 / T_2^2} \quad (11.86)$$

Here, f_c and f_v are Fermi–Dirac distribution functions which are expressed:

$$f_c(E_{21}) = \frac{1}{1 + \exp\left[\left\{ \frac{m_v^*}{m_c^* + m_v^*} (E_{21} - E_g) - E_{fc} + E_c \right\} \Big/ k_B T \right]} \quad (11.87)$$

$$f_v(E_{21}) = \frac{1}{1 + \exp\left[-\left\{ \frac{m_c^*}{m_c^* + m_v^*} (E_{21} - E_g) - E_{fv} + E_v \right\} \Big/ k_B T \right]} \quad (11.88)$$

where k_B is the Boltzmann constant.

The quasi-Fermi distribution may be written:

$$\int_{E_g}^{\infty} \bar{g}(E_{21}) [f_c(E_{21}) + f_v(E_{21}) - 1] dE_{21} = 0 \quad (11.89)$$

Given that the number of injected electrons and holes should be the same, the gain is determined by the following equation:

$$\bar{N} = \int_{E_g}^{\infty} \bar{g}(E_{21}) [f_c(E_{21}) - f_v(E_{21})] dE_{21} \quad (11.90)$$

We have discussed the origin of optical gains in lasers. It is noted that the spectral gain is sometimes inhomogeneous and strongly affects the laser operation, in particular, dynamics and saturations.

PROBLEMS

11.1. Discuss the Doppler effect and its relation to line broadening.
11.2. Obtain Eq. (11.7).
11.3. Prove Eq. (11.13).

11.4. Derive the expression for hole burning and explain the physical meaning of hole burning.
11.5. Obtain Eqs. (11.34) and (11.34′).
11.6. Explain the physical meaning of the Lamb dip.
11.7. Obtain Eqs. (11.64) and (11.65).
11.8. What kind of relationship exists between χ' and χ''?
11.9. Obtain the gain coefficient G per unit time.
11.10. Show the relation of the gain coefficient G and the spatial gain coefficient g per unit length of a gain medium. Hint: $|\Delta z/\Delta t|_{\Delta t \to 0}$ = speed of light in the medium.
11.11. Consider how we can obtain the quasi-Fermi level from Eq. (11.90).

REFERENCES

1. W. E. Lamb, Jr., *Phys. Rev.* **134,** 1429 (1964).
2. R. L. Barger and J. L. Hall, *Phys. Rev.* **139,** 314 (1969).
3. R. H. Pantell and H. E. Puthoff, *Fundamentals of Quantum Electronics,* Wiley, New York (1969).
4. Y. Suematsu, ed., *Semiconductor Lasers and Integrated Optics,* Ohmsha, Tokyo (1984).
5. M. Yamada and Y. Suematsu, *J. Appl. Phys.* **52,** 2653 (1981).

TWELVE

MODULATION AND LIGHT PULSE GENERATION

In this chapter, modulation and light pulse generation in laser oscillators are discussed. First, the delay time associated with the turn-on of laser oscillation and relaxation oscillations are explained by using rate equations. Following that, some methods of generating light pulses of extremely short time duration, i.e., *Q*-switching and mode locking, are introduced. Lastly, direct intensity modulation of lasers by changing the excitation will be treated.

12.1. DELAY IN LASER OSCILLATION

When a laser medium is excited suddenly by a pulsed excitation source, the laser cannot start to oscillate immediately. The laser begins to oscillate after delay time t_d, the characteristic time it takes for the population to build up to a level corresponding to threshold.

In the rate equations (10.46) and (10.47), let us assume that when $t < 0$, $N^e = 0$, and when $t \geq 0$, $N^e = N^e$ (constant). Because the number of photons is very small before oscillation, we can write $S \ll 1$ and the following equation is obtained:

$$\dot{\bar{N}} + \frac{1}{T_1}(\bar{N} - \bar{N}^e) = 0 \tag{12.1}$$

The solution to this equation is readily written as:

$$\bar{N} = \bar{N}^e(1 - e^{-t/T_1}) \tag{12.2}$$

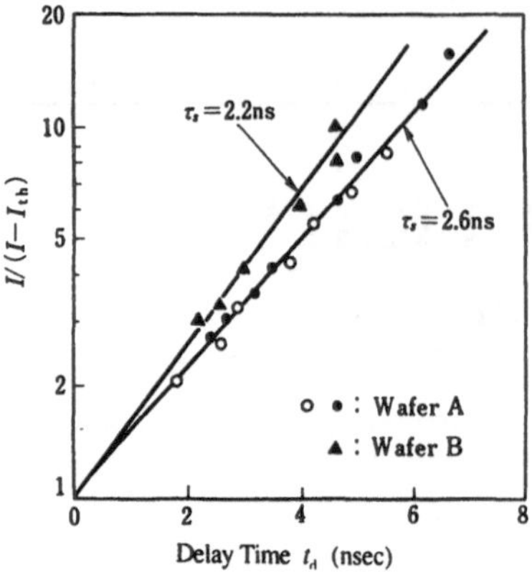

Figure 12.1. Measurements of delay time t_d. ($T_1 = \tau_s$, $N^e/N_{th} = I/I_{th}$.) [After K. Wakao, H. Morita, T. Kambayashi, and K. Iga, *Jpn. J. Appl. Phys.* **16**, 2075 (1977).]

From Eq. (12.2), the time, t_d, required to reach threshold may be expressed as:

$$t_d = T_1 \ln\left(\frac{\bar{N}^e}{\bar{N}^e - 1}\right) = T_1 \ln\left(\frac{N^e}{N^e - N_{th}}\right) \qquad (N_0 = N_{th}) \qquad (12.3)$$

where N_{th} denotes the normalized threshold population difference for oscillation. By changing the excitation level, N^e, and measuring the associated delay times before oscillation, the longitudinal relaxation time, T_1, can be obtained. Figure 12.1 shows an example of this approach where the recombination lifetime τ_s, which corresponds to the longitudinal relaxation time in a semiconductor laser, has been measured.

12.2. RELAXATION OSCILLATION

The delay time of laser oscillation was obtained in the previous section using only the change of population and neglecting the change in the number of photons. Above threshold, the change in the number of photons must be included since stimulated emission becomes important. Thus, to investigate the variation of the number of photons, we use the coupled rate equations. Phenomenologically speaking, with strong excitation, relaxation oscillations occur because the population number density increases until it exceeds the threshold. The number of photons in the cavity then builds up until it exceeds the steady-state value, causing the stimulated emission process to become very large. This strong stimulated emission rate causes the population difference to drop below threshold, and the number of photons rapidly decreases. The rate of stimulated emission then decreases, permitting the population to increase again, and the cycle is repeated. After many such oscillations, the laser gradually reaches a steady-state oscillation condition.

In Section 10.4, the frequency of the relaxation oscillations in response to a small signal excitation was written as:

$$f_r \cong \frac{1}{2\pi T_1}\sqrt{\frac{T_1}{\tau_p}(\bar{N}^e - 1)} \tag{12.4}$$

In the treatment here, the relaxation oscillation mechanism will be explained from another point of view. Rewriting Eqs. (10.46) and (10.47), we have:

$$\frac{d\bar{S}}{dt} = \frac{1}{\tau_p}\bar{S}(\bar{N} - 1) \tag{12.5}$$

$$\frac{d\bar{N}}{dt} = -\frac{\bar{N} - \bar{N}^e}{T_1} + \frac{1}{T_1}(1 - \bar{N}^e)\bar{N}\bar{S} \tag{12.6}$$

If we assume

$$\bar{N} = 1 + \delta$$

and substitute into the right-hand side of Eq. (12.6), letting $\delta \ll 1$, we obtain:

$$\frac{d\bar{N}}{dt} \cong \frac{1}{T_1}(1 - \bar{N}^e)(\bar{S} - 1) \tag{12.7}$$

From the ratio of Eqs. (12.5) and (12.7), we find:

$$\frac{d\bar{S}}{d\bar{N}} = \frac{T_1}{\tau_p} \cdot \frac{(\bar{N} - 1) \cdot \bar{S}}{(1 - \bar{N}^e)(\bar{S} - 1)} \tag{12.8}$$

which leads to:

$$(\bar{N} - 1)d\bar{N} = \frac{\tau_p}{T_1}(1 - \bar{N}^e)\left(1 - \frac{1}{\bar{S}}\right)d\bar{S} \tag{12.9}$$

By integrating both sides, we can write:

$$\frac{1}{2}(\bar{N} - 1)^2 = \frac{\tau_p}{T_1}(1 - \bar{N}^e)\left(\bar{S} - \bar{S}_i - \ln\frac{\bar{S}}{\bar{S}_i}\right) \tag{12.10}$$

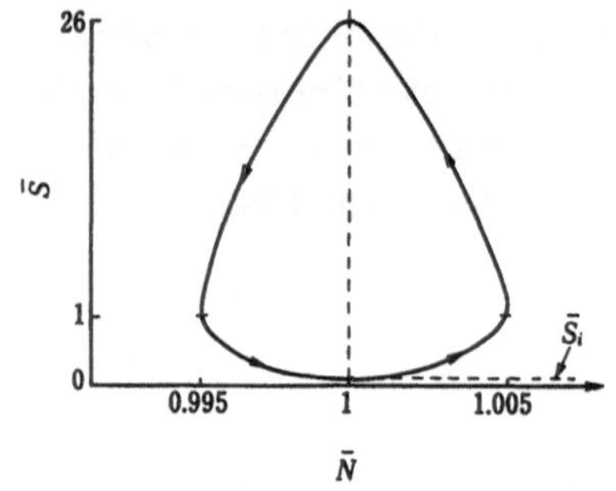

Figure 12.2. Traces of S and N (assuming $S_i = 10^{-10}$). (From Ref. 2.)

where $N = 1$ when $S = S_i$. S_i denotes the initial number of photons generated through spontaneous emission in the resonator normalized to the steady-state photon flux.

Figure 12.2 shows the trace of S and N, assuming $S_i = 10^{-10}$. The locus of points indicates the relaxation oscillation cycle.[2] By assuming the maximum of S to occur at $N = 1$, we can write

$$\bar{S}_{\max} - \bar{S}_i - \ln \frac{\bar{S}_{\max}}{\bar{S}_i} = 0 \tag{12.11}$$

assuming $S_i = 10^{-10}$, we find $S_{\max} = 26$, or a peak photon flux 26 times the steady-state value.

12.3. Q-SWITCHING

From the previous section it is clear that rapid excitation causes a sudden change in the oscillator output and generates strong optical pulses on the order of 26 times the steady-state value. If larger peak outputs are desired, the Q of the resonator has to be reduced to restrain oscillations. This is because in the ordinary case, where the Q factor of the resonator is large, oscillation begins immediately after threshold and quickly suppresses any significant further increase in the population. If the resonator Q is reduced, the population difference may be made as large as possible without being depleted by oscillation. At that point the Q is suddenly switched so that oscillation can start and a giant output pulse can be obtained. This technique is called Q-switching.

In Eq. (12.6), the second term on the right is assumed to dominate since it is responsible for the photon-related dynamics, and S rapidly becomes

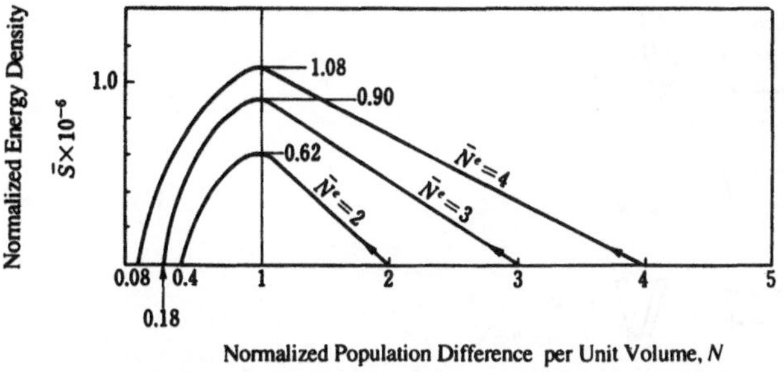

Figure 12.3. Change of output in Q-switching operation. ($T_1/\tau_p = 2 \times 10^6$.) [After R. H. Pantell and H. E. Puthoff, *Fundamentals of Quantum Electronics*, Wiley, New York (1969).]

much, much greater than 1 when the cavity is Q-switched. At that time, the population is assumed to have reached its maximum value. Recall that $\bar{S} = S/S_0$, where S_0 is the steady-state output for the high-Q cavity. Equation (12.6) thus becomes

$$\frac{d\bar{N}}{dt} \cong \frac{1}{T_1}(1 - \bar{N}^e)\,\bar{N}\bar{S} \tag{12.12}$$

From Eqs. (12.5) and (12.12) we find

$$\frac{d\bar{S}}{d\bar{N}} = \frac{T_1}{\tau_p} \cdot \frac{\bar{N} - 1}{(1 - \bar{N}^e)\bar{N}} = -\frac{T_1/\tau_p}{\bar{N}^e - 1}\left(1 - \frac{1}{\bar{N}}\right) \tag{12.13}$$

Integrating the above equation yields:

$$\bar{S} = \frac{T_1/\tau_p}{\bar{N}^e - 1}\left[\ln\frac{\bar{N}}{\bar{N}^e} - (\bar{N} - \bar{N}^e)\right] \tag{12.14}$$

where, as initial conditions, $S = 0$ and $N = N^e$.

From Eq. (12.13), we see that the maximum value of S occurs at $N = 1$. Substituting that into Eq. (12.14) gives

$$\bar{S}_{\max} = \frac{T_1}{\tau_p}\left[1 - \frac{1}{\bar{N}^e - 1}\ln \bar{N}^e\right] \tag{12.15}$$

From this expression it is clear that the maximum output becomes large and is predominantly determined by the ratio T_1/τ_p.

Figure 12.3 shows the normalized energy density versus the normalized population difference for $T_1/\tau_p = 2 \times 10^6$. We see that, for $N^e = 4$, the nor-

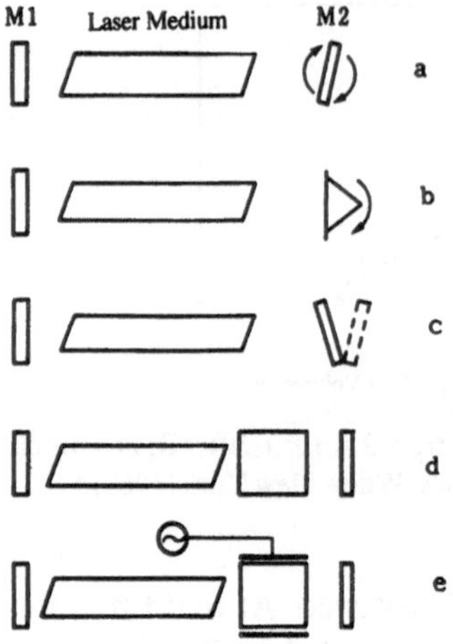

Figure 12.4. Methods of Q-switching operation. (a) Rotating mirror; (b) rotating prism; (c) vibrating mirror; (d) saturable absorber; (e) optical modulator.

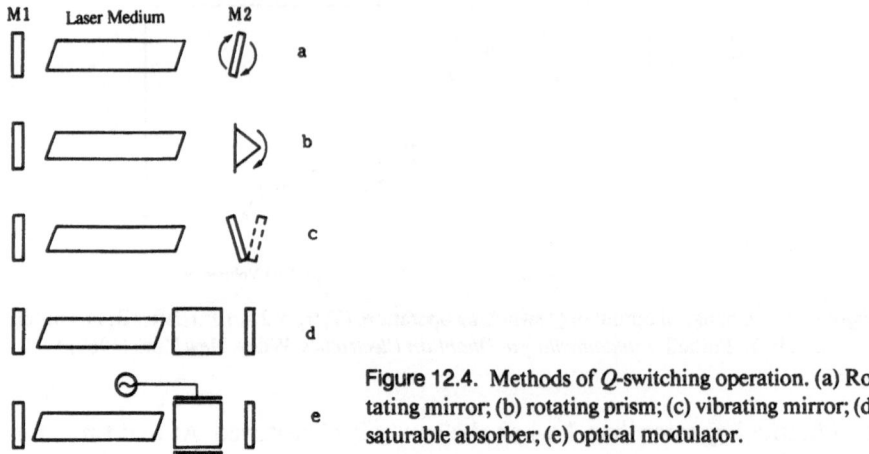

malized energy density can reach a factor as large as 10^6 times the steady-state value.[2]

Figure 12.4 shows several methods of Q-switching. For example, in a Nd:YAG laser, a continuous output on the order of 30 W is available, whereas outputs on the order of 10 MW can be obtained by Q-switching using method (e). Of the various methods shown in Fig. 12.4, only method (d), saturable absorption, is passive. Therefore, Q-switching operation by method (d) cannot be externally controlled.

12.4. MODE LOCKING

Q-switching operation generates optical pulses with large peak values and pulse widths on the order of nanoseconds. In this section, we discuss mode locking, which is a common technique used to generate optical pulses with a duration on the order of picoseconds to femtoseconds.

A single pulse oscillating in an optical cavity has a round-trip time in-

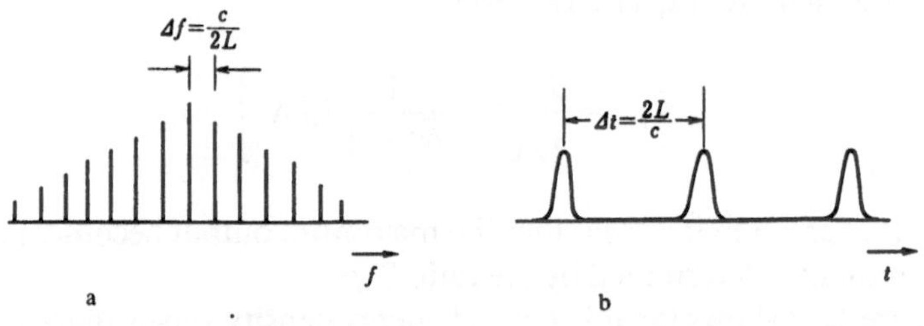

Figure 12.5. Mode locking. (a) Frequency spectrum; (b) pulse waveform.

terval of $2L/c$. This time interval leads to a characteristic frequency, and the Fourier spectrum of this pulse is shown in Fig. 12.5a where the frequency modes occur at intervals of $c/2L$. In a standard oscillator with no mode locking, various modes exist and have a spectrum similar to that shown in Fig. 12.5a. However, the phase of each mode, in general, has no correlation with other modes and, therefore, each oscillates independently. If, by a certain method, the phase relationship of the modes are locked, such as by pulse modulation as shown in Fig. 12.5b, the sum of the multiple frequency modes add up to create a circulating pulse.

Figure 12.6 shows one method of mode locking. An optical switch is placed in the optical resonator. In order to create pulsed oscillation, we have only to switch at time intervals of $2L/c$, corresponding to the round-trip time for the pulse in the resonator. For example, if $L = 1$ m, then the round-trip time is 6×10^{-9} sec and a high-speed switch is required which may, for example, consist of an optical deflector driven by means of ultrasonic diffraction. Such active mode locking in a YAG laser leads to pulses with widths on the order of 100 psec. Shorter pulses can be obtained using passive mode locking where a bleachable dye serves as an optically activated high-speed shutter. These lead to pulses on the order of several picoseconds or less in dye lasers, or tens of picoseconds in Nd:YAG lasers.

The amplitude of the electric field, E, in the optical cavity may be expressed as a sum of the modes E_n, each with angular frequency ω_n and phase ϕ_n:

$$E = \sum_n E_n \exp\left[j\left\{\omega_n\left(t - \frac{z}{c}\right) + \phi_n\right\}\right] \tag{12.16}$$

where the wave propagates in the positive z direction.

If the angular frequency interval between the modes is written as:

$$\Delta\omega = 2\pi \cdot \frac{c}{2L} \tag{12.17}$$

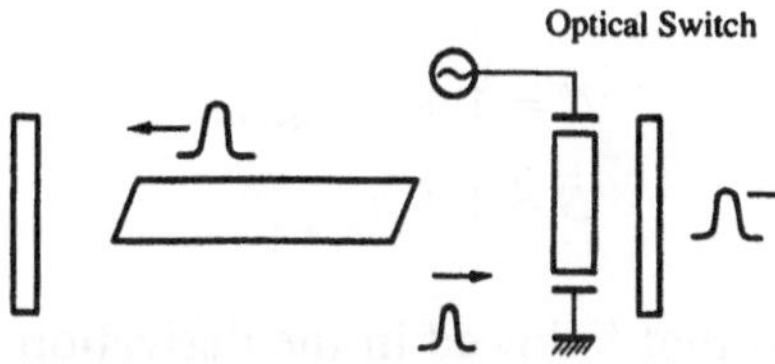

Figure 12.6. Methods of mode locking.

then we find

$$\omega_n = \omega_0 + n\Delta\omega \qquad (n = 1, 2, \ldots) \tag{12.18}$$

From Eq. (12.16), the following is obtained:

$$E = \exp\left\{j\omega_0\left(t - \frac{z}{c}\right)\right\}\left[\sum_n E_n \exp\left\{jn\Delta\omega\left(t - \frac{z}{c}\right) + j\phi_n\right\}\right] \tag{12.19}$$

If the phases ϕ_n are not randomly varying, this expression leads to a mode-locked pulse. Figure 12.5 shows that the resulting wave with envelopes which are determined by the amplitude E_n and phase ϕ_n propagates at the speed of light. The pulse length is determined by the Fourier transform of E and is inversely proportional to the number of modes oscillating in the cavity:

$$\Delta t = \frac{2\pi}{n\Delta\omega} = \frac{2L}{nc} \tag{12.20}$$

12.5. DIRECT MODULATION

The method of modulating the light output of a laser by temporally changing the intensity of the excitation is called direct modulation. On the other hand, the method of modulating by using an optical modulator placed outside of the laser resonator is called external modulation. This section will describe direct modulation, focusing especially on its frequency characteristics.

We can describe direct modulation by assuming a small variation is driven in the equilibrium population difference:

$$\bar{N}^e = \bar{N}_0^e[1 + h \cdot r(t)] \tag{12.21}$$

where N_0^e is assumed to be constant, and $h \ll 1$. This variation then leads to a small change in the population and a small change in the photon flux:

$$\bar{N} = 1 + h \cdot q(t) \tag{12.22}$$

$$\bar{S} = 1 + h \cdot p(t) \tag{12.23}$$

In a manner similar to that followed in the derivation of Eqs. (10.50) and (10.51), we can write the rate equations to terms of p, q, and r:

$$\dot{p} = \frac{1}{\tau_p} q \tag{12.24}$$

$$\dot{q} + \frac{\bar{N}_0^e}{T_1} q + \frac{1}{T_1}(\bar{N}_0^e - 1)\, p = \frac{\bar{N}_0^e}{T_1} r \tag{12.25}$$

Substituting Eq. (12.24) into (12.25) gives:

$$\ddot{p} + \frac{\bar{N}_0^e}{T_1}\dot{p} + \frac{1}{T_1 \tau_p}(\bar{N}_0^e - 1)p = \frac{\bar{N}_0^e}{T_1 \tau_p} r \tag{12.26}$$

Assuming that the modulation is sinusoidal, we have:

$$\left.\begin{aligned} r(t) &= r_1 \exp(j\omega_m t) \\ q(t) &= q_1 \exp(j\omega_m t) \\ p(t) &= p_1 \exp(j\omega_m t) \end{aligned}\right\} \tag{12.27}$$

Substituting these equations into Eq. (12.26) leads to the expression:

$$p_1\left[-\omega_m^2 + j\omega_m \frac{\bar{N}_0^e}{T_1} + \frac{1}{T_1 \tau_p}(\bar{N}_0^e - 1)\right] = \frac{\bar{N}_0^e}{T_1 \tau_p} r_1 \tag{12.28}$$

The modulation efficiency is defined as the ratio of the normalized photon flux variation divided by the normalized driven population variation:

$$M = \frac{p_1}{r_1} = -\frac{\dfrac{\bar{N}_0^e}{T_1 \tau_p}}{\omega_m^2 - \omega_r^2 - j\omega_m / T} \tag{12.29}$$

where

$$\omega_r = \frac{1}{T_1}\sqrt{\frac{T_1}{\tau_p}(\bar{N}_0^e - 1)} \tag{12.30}$$

$$\frac{1}{T} = \frac{\bar{N}_0^e}{T_1} \tag{12.31}$$

$$|M(\omega_m)| = \frac{\bar{N}_0^e/(T_1 \tau_p)}{\sqrt{[\omega_m^2 - \omega_r^2]^2 + \omega_m^2/T^2}} \tag{12.32}$$

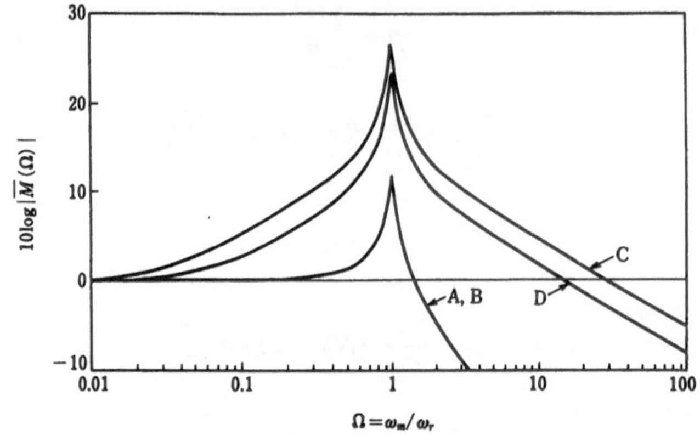

Figure 12.7. Modulation characteristics of laser intensity. A, injection modulation; B, carrier loss modulation; C, gain modulation; D, cavity Q modulation. We have assumed that $P_0/P_{th} = 2$, $\tau_{s0}/\tau_{p0} = 900$, and $\omega_r T = 15$. [After K. Iga, *Trans. IECE Jpn.* **68**, 417 (1985).]

If the modulation efficiency is normalized by the DC value ($\omega_m = 0$), we can write:

$$|\bar{M}(\omega_m)| = \left|\frac{M(\omega_m)}{M(0)}\right| = \frac{\omega_r^2}{\sqrt{(\omega_m^2 - \omega_r^2)^2 + \omega_m^2/T^2}}$$

$$= \frac{1}{\sqrt{\left\{\left(\dfrac{\omega_m}{\omega_r}\right)^2 - 1\right\}^2 + \left(\dfrac{\omega_m}{\omega_r}\right)^2\left(\dfrac{1}{\omega_r T}\right)^2}} \qquad (12.33)$$

The modulation efficiency reaches its maximum value when $\omega_m = \omega_r$. The associated frequency, $f_r = \omega_r/2\pi$, is called a modulation resonance frequency. These relationships were found by Ikegami and Suematsu in 1967.[3] Line A in Fig. 12.7 shows the modulation efficiency, where $\Omega = \omega_m/\omega_r$, $|M(1)|$. Lines B, C, and D are modulation characteristics associated with the modulation of other parameters such as τ_s, τ_p, and the gain constant.

PROBLEMS

12.1. Obtain the delay time of laser oscillation after onset of pumping [Eq. (12.3)].
12.2. Obtain the recombination lifetime τ_s from the delay time t_d. Use the data of Fig. 12.1.
12.3. Obtain Eq. (12.10).
12.4. From Eq. (12.11), obtain $S_{max} = 26$ for $S_i = 10^{-10}$ and for $S_i = 10^{-11}$.

12.5. Obtain Eq. (12.13).
12.6. Prove Eq. (12.14).
12.7. Obtain S_{max} of Eq. (12.15) when $T_1 = 1$ nsec, $\tau_p = 1$ psec, $N^e = 5$.
12.8. Explain the physical meaning of mode locking.
12.9. Discuss the pulse width and linewidth of mode-locked lasers.
12.10. Obtain Eq. (12.29).
12.11. Plot the modulation efficiency M of Eq. (12.33) against the modulation frequency f_m.

REFERENCES

1. K. Wakao, H. Morita, T. Kambayashi, and K. Iga, *Jpn. J. Appl. Phys.* **16,** 2075 (1977).
2. R. H. Pantell and H. E. Puthoff, *Fundamentals of Quantum Electronics,* Wiley, New York (1969).
3. T. Ikegami and Y. Suematsu, *Proc. IEEE* **55,** 122 (Jan. 1967 12).
4. K. Iga, *Trans. IECE Jpn.* **68,** 417 (1985).

THIRTEEN

LASER NOISE

A laser is an oscillator, which generates electromagnetic waves with an optical frequency. The laser, however, does not always emit a continuously oscillating lightwave with a single frequency; its intensity and frequency sometimes fluctuate. These fluctuations are called intensity noise and frequency noise,* respectively. Laser noise is caused by a component spontaneously emitted at random which is mixed with the steady-state optical electromagnetic field. The frequency noise can be expressed by the well-known Schawlow–Townes relation, which represents the linewidth of gas and solid-state lasers. Semiconductor lasers exhibit an index variation resulting from carrier fluctuations and the equivalent length of the laser resonator also fluctuates. This enhances the frequency noise and requires modification of the Schawlow–Townes relation. In this chapter laser noise, which is very important for practical use, will be discussed. Our discussion will take account of the special characteristics of semiconductor lasers.

13.1. INTENSITY NOISE

13.1.1. Measure of Intensity Noise

The light intensity obtained from a continuously operating laser is almost constant. Strictly speaking, however, it fluctuates slightly around the

* Noise originally referred to a sound wave mixed with a signal, and means "shouting voices" in French.

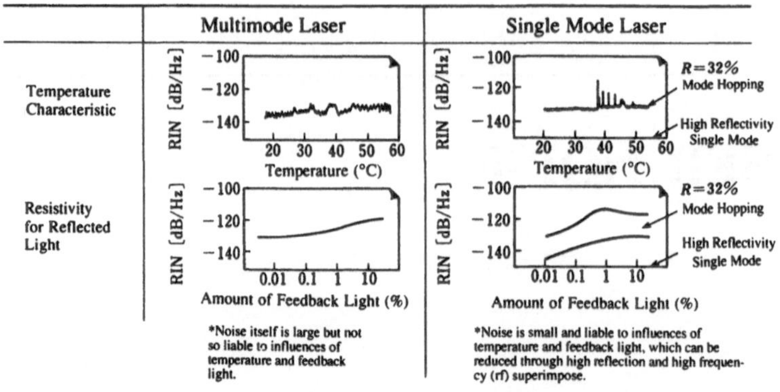

Figure 13.1. Relative Intensity Noise (RIN) of semiconductor lasers.

mean value. This fluctuation is called intensity noise. Let us define relative intensity noise (RIN) to express the intensity noise quantitatively as follows[1,2]:

$$\mathrm{RIN} = [\langle \delta P^2 \rangle / P_0^2]/\Delta f \tag{13.1}$$

or in a decibel unit as

$$\mathrm{RIN} = 10 \log\{\langle \delta P^2 \rangle / P_0^2\}/\Delta f \tag{13.1'}$$

where $\langle \delta P^2 \rangle$ denotes the mean square value of the intensity fluctuation, P_0 denotes the mean optical power, and Δf is the bandwidth near the Fourier frequency f, where we perform the measurement. Figure 13.1 shows RIN of semiconductor lasers.[1,2] As will be mentioned later, in addition to the quantum noise by spontaneous emission, intensity noise is enhanced because of mode hopping caused by temperature change or when a laser beam returns to the laser cavity after being reflected outside the laser. A desirable intensity noise RIN is −140 dB/Hz or less for practical use such as in laser disks.

13.1.2. Quantum Noise

Even if environmental factors around the laser oscillator are constant, some amount of fluctuation still exists. It is caused by quantum noise resulting from spontaneous emission. To describe these effects we have to include quantum noise sources into the rate equations. Instead of the population difference N described in Chapter 10, we use N to express the number of electrons in the conduction band. Then, the rate equations for the photon density S and the carrier density N per unit active volume are given by[1–4]

$$\frac{dS}{dt} = \xi A(N - N_g)S - \frac{S}{\tau_p} + C\frac{N}{\tau_s} + f_s \tag{13.2a}$$

$$\frac{dN}{dt} = -\xi A(N - N_g)S + \frac{N}{\tau_s} + \frac{I}{eV} + f_n \tag{13.2b}$$

where V is the active region volume, I is the injection current, e is the electron charge, the constant A is a differential gain, and ξ is the confinement factor which is defined by the overlap integral of gain and resonant mode intensity. The parameter N_g is a transparency carrier density, above which the semiconductor exhibits positive gain.

Here, we have to note that the factor "2" in Eq. (10.22) has been omitted, since the number of electrons in the conduction band should be equal to the number of holes in the valence band as a result of charge neutrality. In other words, the injection of one electron automatically causes the removal of one electron in the valence band (injection of one hole). One electron emits one photon by recombining with one hole.

The third term on the right-hand side of Eq. (13.2a) expresses the photon generation rate resulting from spontaneous emission which couples to the lasing mode.[1,2] The factor C is a so-called "spontaneous emission factor," sometimes denoted by β, and describes the probability of spontaneously emitted photons contributing to the lasing mode.[5]

We have also introduced f_s and f_n, which represent noise source terms for photons and carriers expressing quantum fluctuations. The auto- and cross-correlations of their Fourier components F_s and F_n are given by the following shot noise expressions[4]:

$$\langle F_s^2 \rangle = 2CVN_0(VS_0 + 1)^3/\tau_s \tag{13.3a}$$

$$\langle F_n^2 \rangle = 2[CV^2N_0S_0/\tau_s + VN_0/\tau_s] \tag{13.3b}$$

$$\langle F_sF_n \rangle = -2CVN_0(VS_0 + 1)/\tau_s + VS_0/\tau_p \tag{13.3c}$$

where $\langle\ \rangle$ is an ensemble average.* S_0 and N_0 denote the densities of photons and carriers under steady-state conditions above threshold and are given by:

* Some errors in Ref. 4 have been corrected.

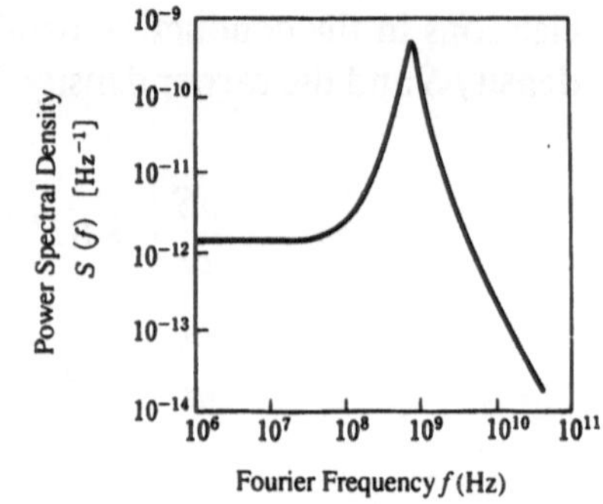

Figure 13.2. Power spectral intensity of IM noise. [After Y. Yamamoto, S. Saito, and T. Mukai, *IEEE J. Quantum Electron.* **QE-19,** 47 (1983).]

$$S_0 = \frac{\tau_p}{\tau_s} N_0(I/I_{\text{th}} - 1) \tag{13.4}$$

$$N_0 = \frac{1}{\xi A \tau_p} + N_g \tag{13.5}$$

where G_{th} in Chapter 10 is replaced by ξA, which expresses the gain coefficient at threshold. Here, we have used τ_s for the longitudinal relaxation time T_1 in Eq. (10.42).

The noise power spectrum $s(f)$ of the photon density as a function of Fourier frequency f is written using the rate equation as[1]

$$\langle [s(f)]^2 \rangle = \frac{a + bf^2}{(f^2 - f_r^2)^2 + \Gamma^2 f^2} \tag{13.6}$$

where f_r denotes the resonant frequency introduced in Chapter 12 modified by including the transparency carrier density:

$$f_r = \frac{1}{2\pi} \sqrt{\frac{1}{\tau_p \tau_s}(1 + \xi A N_g \tau_p)(I/I_{\text{th}} - 1)} \tag{13.7}$$

Here, a, b, and Γ are constants which can be expressed in terms of Eqs. (13.3a)–(13.3c):

$$a = \tau_p^{\text{p}} f_r^4 \langle F_n^2 \rangle + \Gamma^2 \langle F_s^2 \rangle/(2\pi)^2 + 2\tau_p f_r^2 \Gamma \langle F_n F_s \rangle/(2\pi) \tag{13.8}$$

$$b = \langle F_s^2 \rangle/(2\pi)^2 \tag{13.9}$$

$$\Gamma = \frac{1}{2\pi}(\xi A S_0 + 1/\tau_s) \tag{13.10}$$

This relationship is almost the same as the modulation characteristic of lasers discussed in Chapter 12. Figure 13.2 shows an example for the frequency characteristics of the intensity noise of a semiconductor laser.[2] The quantum noise is almost constant in the low-frequency region, but peaks at $f = f_r$.

If we do not pay attention to the resonant peak at f_r, we would say that the intensity noise power spectrum at low frequencies is "white." When the frequency exceeds the resonant frequency, RIN decays with $1/f^2$.

Let us now consider the dependence of RIN on the excitation level:

$$R = (I/I_{\text{th}} - 1) \tag{13.11}$$

When we rewrite RIN in terms of photon density instead of powers as in Eq. (13.1), we have

$$\text{RIN} = [\langle s(0)^2\rangle / S_0^2] \tag{13.12}$$

From Eqs. (13.3), (13.8), and (13.9), RIN is given by the following expression[4]:

$$\text{RIN} = 2\tau_p^2\left[\frac{(CN_{\text{th}}/\tau_s)}{(\xi A\tau_s)^2}\cdot\frac{1}{S_0^3} + \left(\frac{N_0}{V\tau_s} + \frac{1}{\xi AV\tau_p\tau_s}\right)\cdot\frac{1}{S_0^2} + \frac{1}{\tau_s}\cdot\frac{1}{VS_0}\right] \tag{13.13}$$

The first term relating to the beat component between signal and spontaneous emission can be expressed for the range $0.01 < R < 1$ when $\xi A\tau_p N_g \ll 1$ as:

$$\text{RIN} = 2\tau p^2\frac{(CN_{\text{th}}/\tau_s)}{(\xi A\tau_s)^2}\cdot\frac{1}{S_0^3} = 2\tau_p C/(I/I_{\text{th}} - 1)^3 = 2\tau_p C/R^3 \tag{13.14}$$

The derivation of this equation requires some tedious calculation, but the final result is very simple. Note that RIN related to beat noise from spontaneous emission decreases inversely proportional to the cubic power of the drive level $[R^3 = (I/I_{\text{th}} - 1)^3]$.

13.1.3. Enhancement of Intensity Noise by Various Factors

In addition to spontaneous emission, some external factors may enhance the noise fluctuations. Furthermore, in the case where a large number of longitudinal modes of the laser exist, mode hopping occurs, also resulting in large noise. In some cases, the longitudinal modes averaged out to make the noise effect small. The various factors of intensity noise are summarized below:

- Change in output power caused by temperature variation
- Change of injection current
- Instability of laser oscillation resulting from reflection
- Mode and polarization hopping
- Shot noise of photodetectors used to measure the output power of lasers ($\propto 1/R$)

13.1.4. Stabilization of Laser Output

The fluctuation of the laser output resulting from temperature changes is normally slow. Therefore, to obtain stable output, an automatic power control (APC) is utilized which monitors the output and feeds back a stabilization signal.

13.2. FREQUENCY NOISE

13.2.1. Expression for Frequency Noise

The frequency of a laser is not always constant, but fluctuates around a mean optical frequency ν_0. The optical frequency fluctuation $\Delta\nu$ is written as:

$$\Delta\nu = \sqrt{\langle(\nu - \nu_0)^2\rangle} \tag{13.15}$$

where $\langle\ \ \rangle$ denotes the ensemble average. As mentioned later, we should be careful when this average is calculated in order to standardize the mean value.

13.2.2. Quantum Noise

Fluctuations $\Delta\nu$ of the optical frequency are also caused by spontaneous emission. Spontaneously emitted light with random phase is mixed with the light oscillating in a steady state and diffuses the phase, which results in a degradation of coherence. This frequency fluctuation was mathematically expressed by Schawlow and Townes. The formula is known as the Schawlow–Townes relation[5] and given by

$$\delta\nu_{ST} = 8\pi h\nu(\Delta\nu_c)^2/P \tag{13.16}$$

Here, $h\nu$, $\Delta\nu_c$, and P denote the photon energy, resonance width of a laser resonator, and output power, respectively.

In a semiconductor laser the linewidth is broadened by $(1 + \alpha^2)$ in addition to the Schawlow–Townes relation.[6] This should be called a modified Schawlow–Townes relation, and written as:

$$\Delta\nu_{MST} = 8\pi h\nu(\Delta\nu)^2\,(1 + \alpha^2)/P \tag{13.17}$$

Here,

$$\alpha = \Delta n'/\Delta n'' \tag{13.18}$$

is a parameter called a linewidth enhancement factor which was defined by C. Henry.[7]

In the above, the change of refractive index in a semiconductor was written as

$$\Delta n = \Delta n' - j\Delta n'' \tag{13.19}$$

This relationship is obtained as follows: The electric field $\phi(t)$ oscillating with an angular frequency ω_0 can be written as

$$\phi(t) = \phi_0 \exp(j\omega_0 t) \tag{13.20}$$

This obeys the following differential equation:

$$d\phi/dt = j\omega_0\phi \tag{13.21}$$

If we have a change in net gain ΔG and an external force F_ϕ to perturb the field, we have

$$d\phi/dt = j(\omega_0 + \Delta\omega_0)\phi + F_\phi \tag{13.22}$$

where

$$\Delta\omega_0 = (1/2)\alpha\Delta G \tag{13.23}$$

The fluctuation of the net gain ΔG is already given in Eq. (13.2a) and written as:

$$\Delta G = A(N - N_g) - \frac{1}{\tau_p} \tag{13.24}$$

which is the difference of optical gain and loss. The parameter α has also been defined by Eq. (13.18).

The last term in Eq. (13.22) is a quantum fluctuation by spontaneous emission. The average value of its Fourier component may be expressed with the help of Eq. (13.3a) and is given by

$$\langle |F_\phi|^2 \rangle \propto (1/2) \langle |F_s|^2 \rangle \tag{13.25}$$

The frequency fluctuation $\Delta\nu$

$$\Delta\nu = (1/2\pi)\sqrt{\langle(\omega - \omega_0)^2\rangle} \tag{13.26}$$

can be estimated by solving Eq. (13.22) together with Eq. (13.2b), by which Eq. (13.17) is obtained.

In terms of the common semiconductor laser parameters, we get the following equation[7]:

$$\Delta\nu_{\mathrm{MST}} = \frac{h\nu n_{\mathrm{sp}}}{8\pi}\left(\frac{c}{n_1 L}\right) 2\left[\alpha_{\mathrm{a}} + \frac{1}{2L}\ln\left(\frac{1}{R_1 R_2}\right)\right]\ln\left(\frac{1}{R_1}\right)(1 + \alpha^2)/P \tag{13.27}$$

In Eq. (13.27) we have defined the parameters as follows:

c velocity of light
L length of the resonator
α_{a} optical loss in the laser
n_{sp} probability in spontaneous emission
n_1 equivalent refractive index
R_1 reflectivity of the output reflecting mirror
R_2 reflectivity of the rear side reflecting mirror

In the semiconductor laser, the number of carriers changes because of

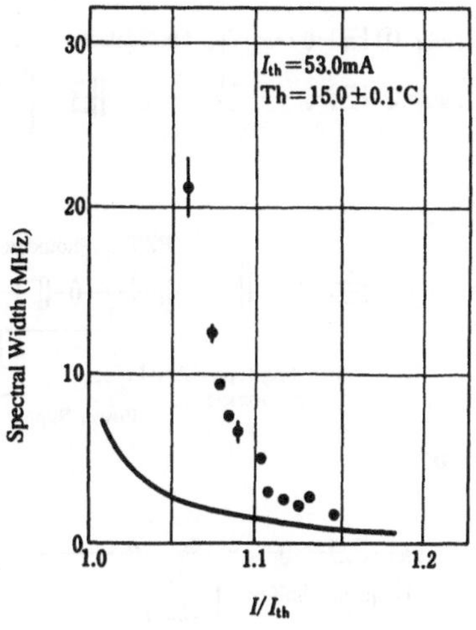

Figure 13.3. Measurement of the frequency width of a semiconductor laser. [After T. Takakura and K. Tako, *Jpn. J. Appl. Phys.* **19,** L725 (1980).]

spontaneous emission, which causes the refractive index of the medium to change. Figure 13.3 shows the measured linewidth of a semiconductor laser.[8]

13.2.3. Enhancement of Frequency Noise by Various Factors

Figure 13.4 shows the power spectrum of the frequency noise of a semiconductor laser,[3] where we note nearly white noise due to spontaneous emission, a resonant peak due to the carrier change peculiar to semiconductor lasers, and other external influences.

13.2.4. Measuring Linewidth

In the case where the laser frequency changes at a relatively slow rate, measurement is easy. One can observe it with a spectrometer. However, if

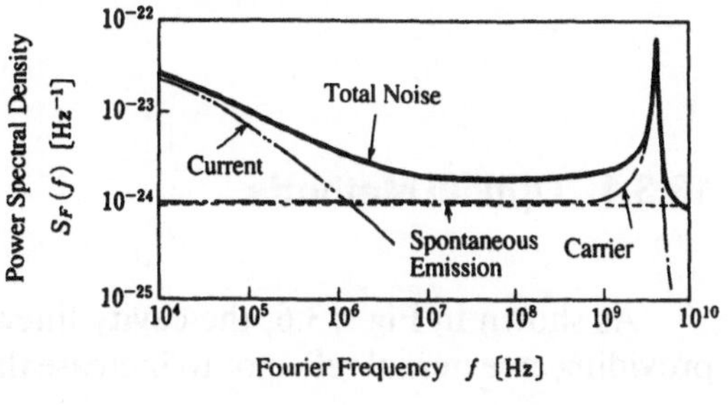

Figure 13.4. Power spectral density of FM noise. [After Y. Yamamoto, S. Saito, and T. Mukai, *IEEE J. Quantum Electron.* **QE-19,** 47 (1983).]

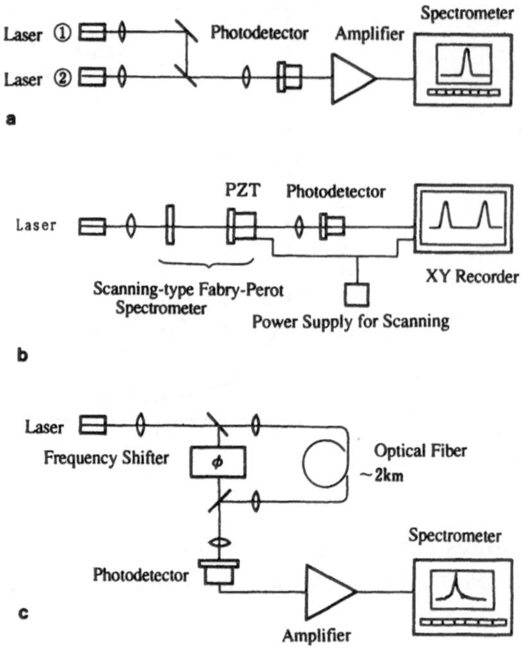

Figure 13.5. Measuring the laser linewidth. (a) Beat measurement between lasers; (b) measurement by using a Fabry–Perot spectrometer[8]; (c) delay self-heterodyne method.[9]

the noise changes randomly and by small amounts, appropriate methods need to be used. One is shown in Fig. 13.5. In order to measure the linewidth with good sensitivity, the delayed homodyne method (see Fig. 13.5c) is employed. This uses an optical fiber and observation of an interference signal.[9]

13.3. CONTROL OF LINEWIDTH

The laser linewidth must be as narrow as possible in cases where the coherence of light is utilized. In some cases, control systems are adopted. Some methods for controlling the linewidth are as follows.

13.3.1. Optical Methods

As shown in Fig. 13.6, the cavity linewidth $\Delta\nu_c$ can be made small by providing an external reflector to increase the Q value of the resonator.[10]

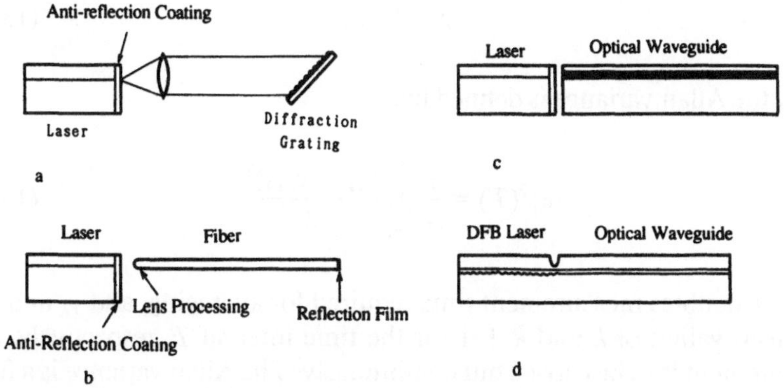

Figure 13.6. Methods for optically narrowing the laser linewidth. (a) External mirror, diffraction grating; (b) optical fiber loading; (c) optical waveguide loading; (d) monolithic-type.

13.3.2. An Electric Method

An electric method[11] is a method where the linewidth of the laser is controlled by an electric feedback following detection of the frequency fluctuation as shown in Fig. 13.7.

13.4. LASER FREQUENCY STABILIZATION

13.4.1. Allan Variance

To quantitatively evaluate the fluctuation of the laser frequency, a mathematical algorithm is needed. The Allan variance introduced by D. Allan is a well-known variance expression.[12] By assuming that ν_0 and $\Delta\nu$ are the nominal frequency and its fluctuation, respectively, let us define the normalized frequency y as:

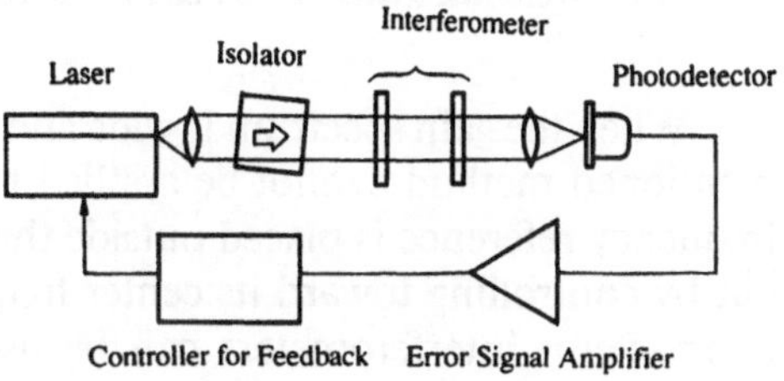

Figure 13.7. Control of linewidth of the laser by electric feedback. [After M. Ohtsu, M. Hashimoto, and H. Ozawa, *Proc. 39th Annu. Symp. Freq. Contr.* (1985).]

$$y = \Delta\nu/\nu_0 \tag{13.28}$$

Then, the Allan variance is defined by:

$$\sigma_y^2(T) = \frac{1}{N}\sum_{k=1}^{N}\frac{(y_k - y_{k+1})^2}{2} \tag{13.29}$$

where T denotes measurement time required for averaging, and y_k and y_{k+1} are mean values of k and $k + 1$ for the time interval T, respectively. The measurement is to be carried out continuously. The Allan variance is a function of the averaging time T, and the number of samples N to obtain the variance must be sufficiently large.

13.4.2. Stabilization at the Center of Laser Gain

In a gas laser, the gain spectrum width is relatively narrow. Therefore, it is practical to control the oscillation frequency at the center of gain spectrum. However, as mentioned in Chapter 8, a narrow dip called a Lamb dip is generated at the center of the output spectrum of a gas laser. A more sensitive control method is, therefore, to match the frequency to the Lamb dip center. When a molecular cell having an absorption spectrum with a narrower width is placed in the laser resonator, saturation of the absorption occurs and a very sharp peak at the center called an inverted Lamb dip is generated. By controlling the frequency toward this center, very high stability can be obtained. The utilization of a methane (CH_4) cell together with 3.39-μm He–Ne laser was proposed by Shimoda[13] and experiments were carried out worldwide.[14] This is employed as a frequency standard in the optical region.[15]

13.4.3. Stabilization with an External Frequency Standard

When the gain spectrum is wide and affected by temperature, the above-mentioned method cannot be applied to get a good stability. Therefore, a frequency reference is placed outside the cavity, and stabilization is carried out by controlling toward its center frequency. As a frequency standard, a Fabry–Perot interferometers can be used, and some absorption lines of

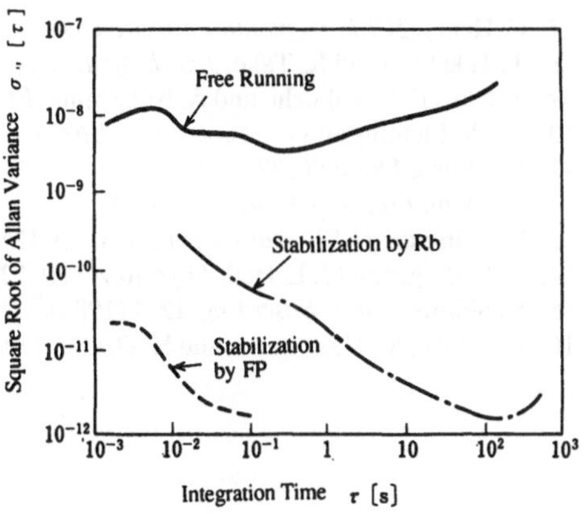

Figure 13.8. Frequency stability of a semiconductor laser. FP, Fabry–Perot interferometer. [After M. Ohtsu, M. Hashimoto, and H. Ozawa, *Proc. 39th Annu. Symp. Freq. Contr.* (1985).]

atoms or molecules are utilized for absolute stabilization. Figure 13.8 shows an example of measured Allan variance in the case where the frequency of a semiconductor laser is stabilized by Rb atoms.[16]

PROBLEMS

13.1. Explain the origin of laser intensity noise.
13.2. Prove Eqs. (13.3) and (13.13).
13.3. Explain mode hopping and its effect on intensity noise.
13.4. Explain why the laser has a finite linewidth.
13.5. By referring to Ref. 7, obtain the linewidth of a semiconductor laser.
13.6. Explain the origin of enhanced broadening of linewidth in semiconductor lasers.
13.7. Summarize the methods for stabilizing the frequency of lasers.
13.8. How can the linewidth of a laser be narrowed?
13.9. We can reduce the linewidth of a laser by electrical feedback of the frequency deviation to control the laser resonator. The linewidth in a well-controlled case can be reduced below the level of the Schawlow–Townes relation. Does this statement violate a physical principle?

REFERENCES

1. H. Haug, *Phys. Rev.* **184,** 338 (1969).
2. Y. Yamamoto, *IEEE J. Quantum Electron.* **QE-19,** 34 (1983).
3. Y. Yamamoto, S. Saito, and T. Mukai, *IEEE J. Quantum Electron.* **QE-19,** 47 (1983).
4. F. Koyama, K. Morito, and K. Iga, *IEEE J. Quantum Electron.* **27,** 1410 (1991).
5. K. Furuya and Y. Suematsu, *Trans. IECE Jpn.* **60,** 467 (1977).
6. A. L. Schawlow and C. H. Townes, *Phys. Rev.* **112,** 1940 (1958).

7. C. Henry, *IEEE J. Quantum Electron.* **QE-18,** 259 (1982).
8. T. Takakura and K. Tako, *Jpn. J. Appl. Phys.* **19,** L725 (1980).
9. T. Okoshi, K. Kikuchi, and A. Nakayama, *Electron. Lett.* **16,** 630 (1980).
10. M. W. Fleming and A. Mooradian, *IEEE J. Quantum Electron.* **QE-17,** 44 (1981).
11. M. Ohtsu, *Physics* **6,** 297 (1985).
12. D. Allan, *Proc. IEEE* **54,** 221 (1966).
13. K. Shimoda, *IEEE Trans. Instrum. Meas.* **IM-17,** 343 (1968).
14. R. L. Barger and J. L. Hall, *Phys. Rev. Lett.* **22,** 4 (1969).
15. K. Shimoda, *Rev. Laser Eng.* **12,** 4 (1984).
16. M. Ohtsu, M. Hashimoto, and H. Ozawa, *Proc. 39th Annu. Symp. Freq. Contr.* (1985).

FOURTEEN

ADVANCED TECHNOLOGY FOR SEMICONDUCTOR LASER FABRICATION AND INTEGRATION

In this chapter we discuss some technology associated with advanced semiconductor lasers, including crystal growth and fabrication methods, laser characterization, longitudinal mode control, system integration, etc. In addition, the integration of semiconductor lasers will be introduced for future lightwave systems.

14.1. METHODS OF SEMICONDUCTOR CRYSTAL GROWTH

14.1.1. Outline of Crystal Growth Method

There are basically two steps of crystal growth, i.e., bulk crystal growth for substrate preparation, and thin-film epitaxy. For a good-quality crystal, the horizontal Bridgeman method is used to produce polycrystal materials with low impurity levels. The crystal is grown from the source prepared at one end of horizontally placed molten metal having a temperature gradient. A liquid encapsulate (LEC) method is used for the growth of big, single crystals which are needed as GaAs or InP substrates. Figure 14.1 shows a schematic diagram of the LEC method. The seed crystal is pulled up with rotation during the growth. Temperature uniformity during crystal growth is very important, and, for this reason, the liquid encapsulate is used to minimize the thermal dissipation by covering the molten metal surface with B_2O_3.

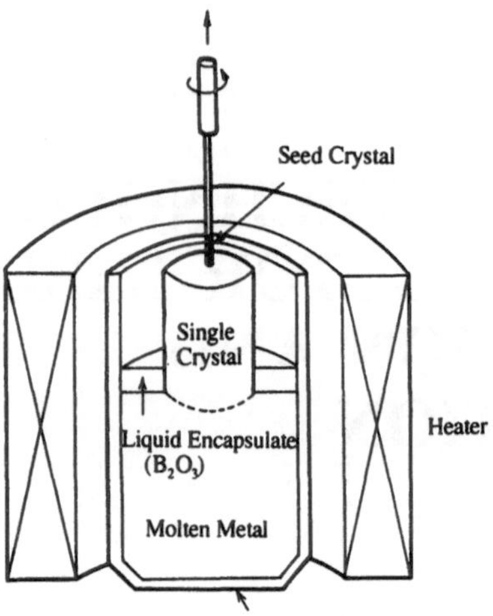

Figure 14.1. Outline of the LEC method.

14.1.2. Liquid-Phase Epitaxy

Liquid-phase epitaxy (LPE), vapor-phase epitaxy, molecular beam epitaxy, and chemical beam epitaxy are used to obtain multilayer wafers for semiconductor lasers. Let us first discuss LPE.[1]

Elements of the desired crystal components, such as As, P, and so on, are melted into a metal solution (Ga in the GaAs system, In in the GaInAsP system), which is kept at typically 650 to 850°C until it is saturated. The solution is then cooled somewhat so that it becomes supersaturated. If a substrate crystal then comes into contact with the solution, the supersaturated material precipitates onto the substrate as a solid phase. To obtain a heterostructure, a slide boat method is utilized.

14.1.2.1. GaInAsP/InP Growth by LPE

Figure 14.2 shows both a schematic diagram of equipment required for LPE[1] and a photograph of a graphite boat rotatable in the furnace. In order to grow multilayers of $Ga_xIn_{1-x}As_yP_{1-y}$, solid-phase precipitation from a supercooled solution and a slide boat are utilized. In this case, In is the metal solvent, and GaAs, InAs, and InP are used as sources of Ga, As, and P. The segregation coefficient, $k_N = x_N^s/x_N^l$, denotes the rate of x_N^s (N = Ga, In, As, and P) precipitated compared with the mole fraction of the liquid-phase component, x_N^l, in the solution. Values of k_N depend on the characteristic

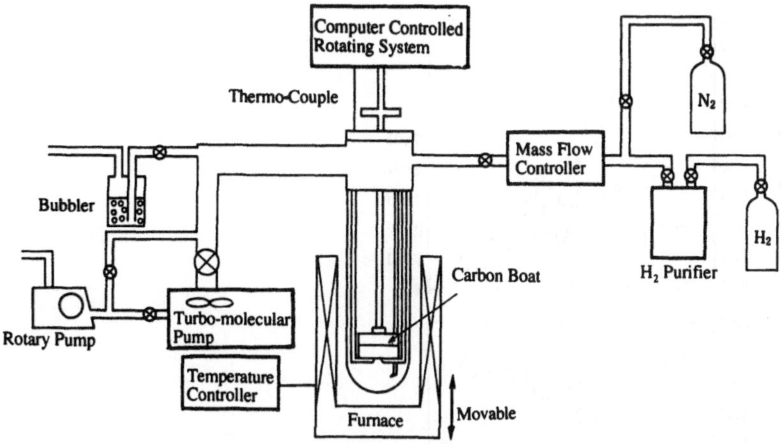

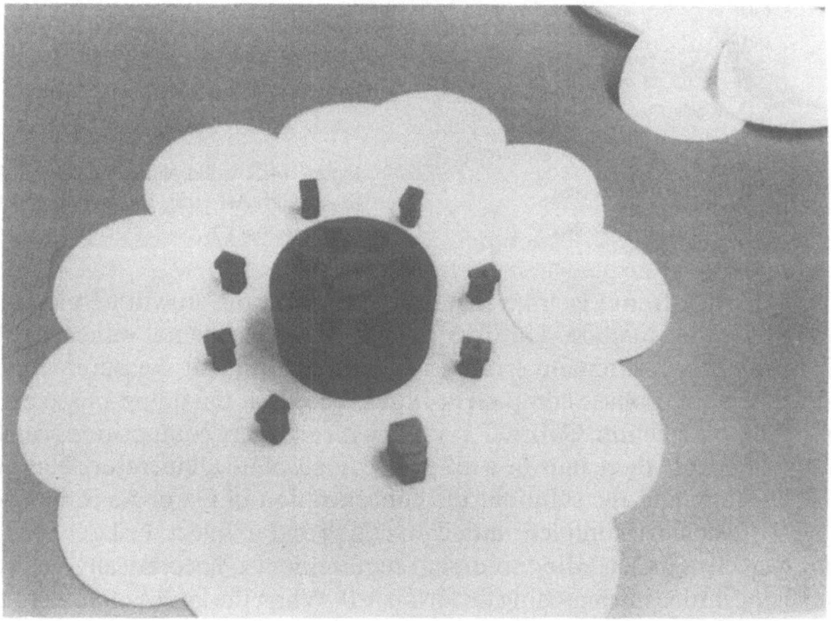

Figure 14.2. a. Example of LPE equipment. b. Photograph of Carbon Boat.

component $x = x^s_{Ga}$ and $y = x^s_{As}$. To understand the parameters of crystal growth, extensive studies were made between 1976 and 1980. The solid-phase component, x^s_N, was obtained experimentally at growth start temperatures between 630 and 650°C through the supercooling method. Figure 14.3 shows the liquid-phase component as a function of the solid-phase component. From this figure the quantity of source crystal that must be dissolved in an In solution can be determined.

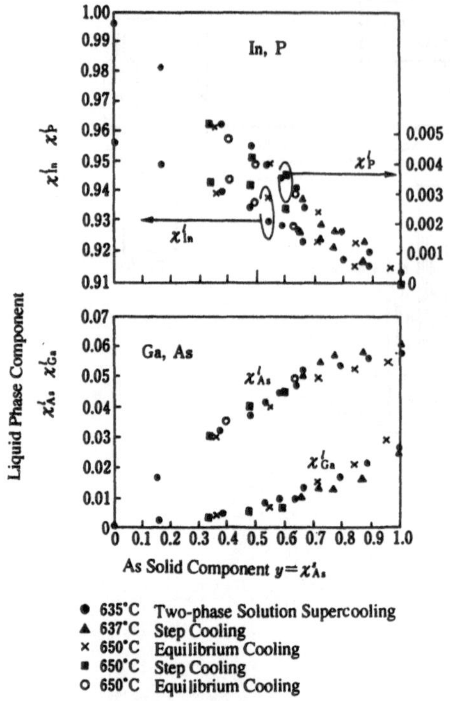

Figure 14.3. Solid-phase component X^{1}_{N} in a GaInAsP/InP semiconductor (N = Ga, In, As, P).

It is clear from Fig. 14.3 that x^{1}_{N} depends on the growth temperature and the cooling method. The dispersion in the experimental values may be the result of measurement errors of the source crystal or measurement errors of the solid-phase component ratios. Based on the liquid-phase component mole fraction, GaInAsP crystals with arbitrary compositions can be grown. Because there may be a difference in absolute temperature between the substrate and the solution, the concentration of Ga or As is changed slightly to achieve complete lattice matching and to select the laser oscillation wavelength according to design requirements. Theoretically, we can make the lattice mismatching factor $\Delta a = 0$. When the lattice of quaternary crystals mismatches that of InP, the distortion of the lattice mainly appears vertically, and there is little distortion in the direction parallel to the substrate surface. For this reason, if the crystal orientation is selected vertically to a substrate of (400) crystal orientation, mismatching can be measured by x-ray diffraction.

At an early stage in research development, the (100) face was used for crystal growth, but small holes of about 1 μm remained on the substrate surface. After a semiconductor laser was successfully used with a (100) substrate, the (100)-oriented substrate came to be commonly used. The (100) face has many advantages, such as easy cleavage and easy epitaxy in selective

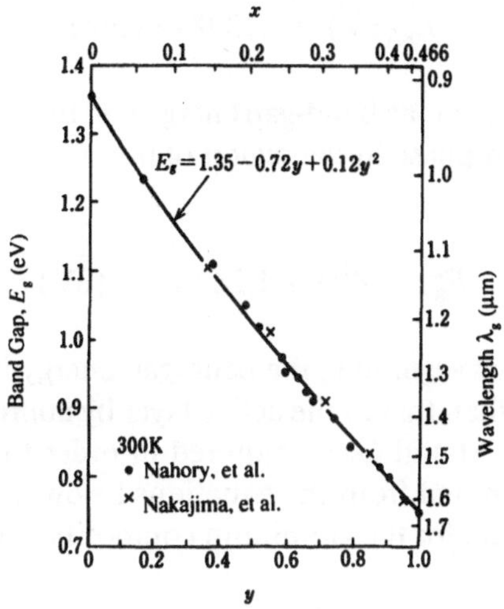

Figure 14.4. E_g and wavelength λ_g versus y.

growth, etching, and lattice matching. In addition, techniques developed for GaAs can be utilized for processing.

The relationship between the solid-phase component ratios of quaternary crystals $x = x^s_{As}$ and $y = x^s_{As}$, and band-gap width, E_g, is very important. This can be determined by substituting experimental values into a so-called Vegard's law. The relationship between x and y is written as:

$$x = \frac{0.466y}{1.03 - 0.03y} \qquad (0 \leqq x \leqq 1) \qquad (14.1)$$

The band gap, E_g, is determined by the solid-phase component ratio y:

$$E_g(y) = 1.35 - 0.72y + 0.12y^2 \qquad \text{(eV)} \qquad (14.2)$$

Figure 14.4 shows E_g and the wavelength, λ_g, plotted as a function of y.

14.1.2.2. GaAlAs/GaAs Growth by LPE

Since the lattice constant of GaAs (5.65325 Å at 27°C) is nearly equal to that of AlAs (5.6605 Å at 0°C) in $Ga_{1-x}Al_xAs$ epitaxy, the band-gap energy can be changed by changing the composition factor, x. If the desired oscillation wavelength is λ_g, the associated band-gap energy of the active layer is given by the conversion expression:

$$E_g\,(\text{eV}) = 1.2398/\lambda\,(\mu\text{m}) \tag{14.3}$$

The relationship between the band-gap energy and the Al composition factor x in $Ga_{1-x}Al_xAs$ solid phase in the region of $0 < x < 0.45$ at 197 K is given by:

$$E_g = 1.424 + 1.247x \qquad (\text{eV}) \tag{14.4}$$

In order to provide waveguiding, the band-gap energy in the cladding layer is usually set higher than that in the active layer by approximately 0.3 eV.

The amount of Al and GaAs required in order to grow a $Ga_{1-x}Al_xAs$ crystal can be determined from the equations below. Here x^{l}_{As} and x^{l}_{Al} are given as measured data for the aluminum composition $x = x^{s}_{Al}$:

$$\frac{x}{M_{Al}} = x^{l}_{Al}\left(\frac{2y}{M_{GaAs}} + \frac{x}{M_{Al}} + \frac{1}{M_{Ga}}\right) \tag{14.5}$$

$$\frac{y}{M_{GaAs}} = x^{l}_{As}\left(\frac{2y}{M_{GaAs}} + \frac{x}{M_{Al}} + \frac{1}{M_{Ga}}\right) \tag{14.5'}$$

where M denotes the atomic or molecular weight. After the desired amounts of Al, GaAs, and Ga are weighed and rinsed in an organic solvent, they are charged into the box of the solution holder. After melting, if they are brought into contact with the GaAs substrate, $Al_xGa_{1-x}As$ single crystals with the desired band-gap energy can be grown.

In order to make a *pn* junction, dopants are mixed into the materials to be charged. In a GaAs system, Ge and Zn are used as p-type dopants and Sn and Te are used as n-type dopants. This type of LPE is still commonly used in production to make wafers for GaInAsP/InP lasers. This approach permits the formation of lasers which operate in the wavelength range of 1.3 to 1.5 μm, and have threshold current densities of 1 kA/cm^2-μm. The lowest laser threshold is approximately 0.4 kA/cm^2, corresponding to a quantum well active layer.

14.1.3. Vapor-Phase Epitaxy

Vapor-phase epitaxy (VPE) and chemical vapor deposition (CVD) both generate crystal growth from materials in a vapor phase. Figure 14.5 is a schematic diagram of VPE equipment reported by Neuse (RCA). NEC[2] and Bell Laboratories developed a method to separate the flows by dividing

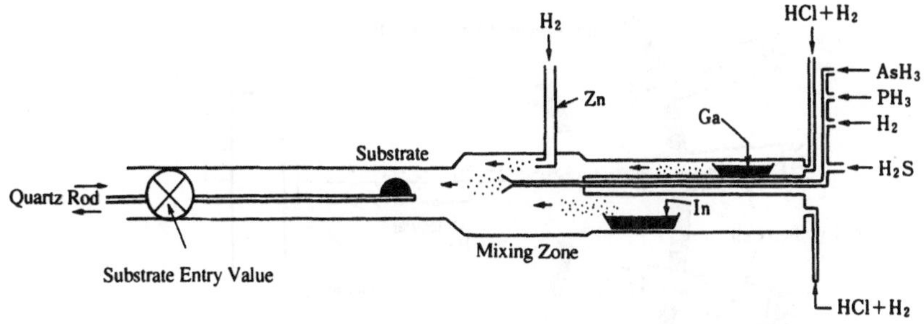

Figure 14.5. Example of VPE equipment. (After Neuse and Olsen.[2])

the reaction tube into two parts.[3] By the close of 1987, the oscillation of a GaInAsP/InP laser grown by VPE was reported at the International Conference on Semiconductor Lasers, where the initial threshold current density of $J_{th} = 7\ \mathrm{kA/cm^2}$-$\mu$m. The VPE approach has been found to be very useful for producing large-sized wafers having good film thickness controllability. These wafers are made predominantly for optical integrated circuits and photodetectors.

14.1.4. Metal–Organic Chemical Vapor Deposition

For metal–organic CVD (MOCVD), elements required for crystal growth are provided through the metal–organic gases. Figure 14.6 presents typical MOCVD equipment.

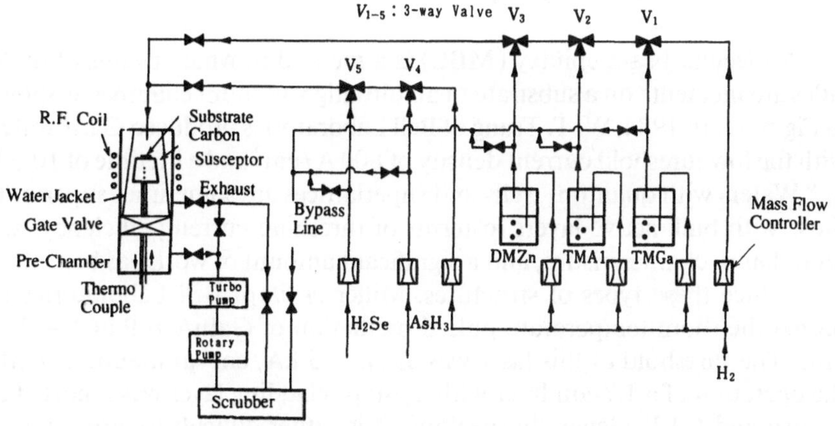

Figure 14.6. Illustrative MOCVD equipment. (After Koyama, Iga *et al.*)

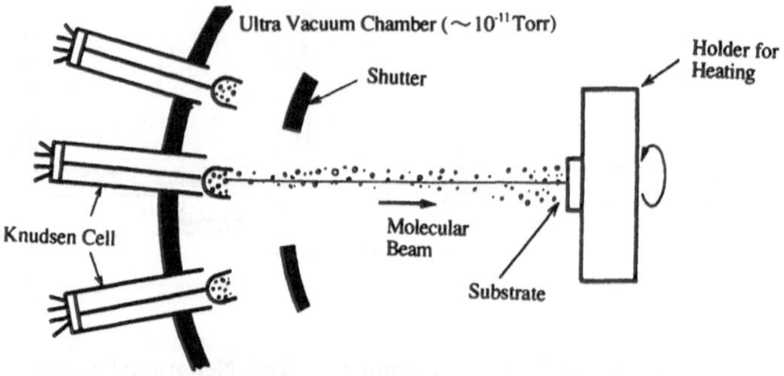

Figure 14.7. Typical MBE equipment.

Dupuis *et al.* of Rockwell[4] were the first to succeed in growing GaAlAs/GaAs lasers by MOCVD. CW operation with a life of 1800 hr was demonstrated in 1979, and the formation of devices with high reliability was reported. A continuous test of 500 hr was carried out by STL in 1979, and the record was increased to 8000 hr in 1980. Lasers with excellent performance are also fabricated by Thomson-CSF, SONY, and Toshiba. Thomson-CSF first reported pulsed oscillation in quaternary GaInAsP laser crystals at 1.15 μm ($J_{th} = 5.9\ kA/cm^2$).[5] After that, crystals of good quality became available that could operate in the region from 1.3 to 1.6 μm. Since 1985, there has been an effort to produce semiconductor lasers such as InGaAlP with visible wavelengths using this method, and lasers in the region from 0.62 to 0.67 μm have become available.

14.1.5. Molecular Beam Epitaxy

Molecular beam epitaxy (MBE) is a method in which beams of molecules are incident[6] on a substrate in an ultrahigh-vacuum chamber as shown in Fig. 14.7. In 1984, W. T. Tsang of Bell Laboratories produced GaAs wafers with the low threshold current density of 800 A/cm^2 and a lifetime of 10,000 hr.[6] Wafers with quantum wells and superlattices are sometimes superior to those with bulk active layers in terms of threshold current, efficiency, and modulation characteristics, and a significant amount of work has been done to produce these types of structures. Miller *et al.* of Bell Laboratories reported the room-temperature pulsed oscillation of GaInAsInP at $\lambda = 1.67$ μm.[7] The threshold of this laser was $J_{th}/d = 5\ kA/cm^2$-μm. Subsequently, the operation of a 1.7-μm laser with a *p*-type cladding layer was reported by Fujitsu and NTT. Generally speaking, it is rather difficult to grow crystals that contain elements for which vapor pressures are very different, for exam-

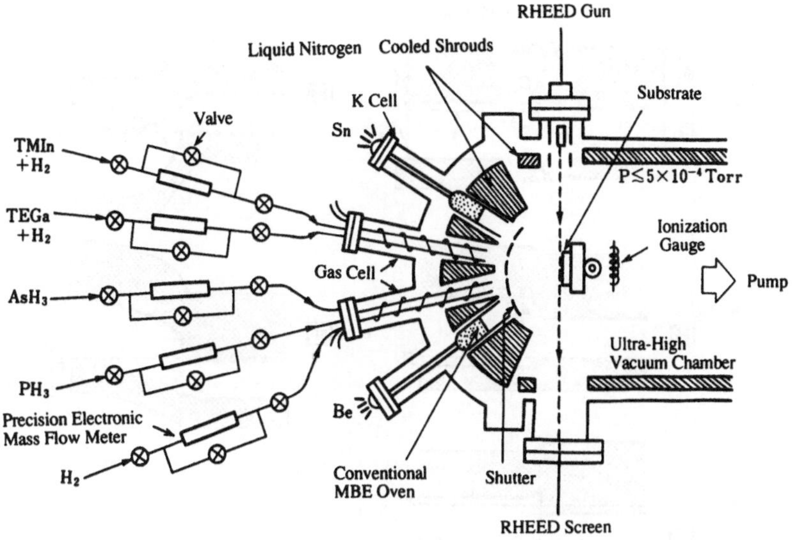

Figure 14.8. Example of CBE equipment. (After Tsang.[8])

ple, arsenic and phosphorus. Therefore, the quaternary crystals are produced by gas source epitaxies.

14.1.6. Chemical Beam Epitaxy

Chemical beam epitaxy (CBE) was conceived by W. T. Tsang of Bell Laboratories.[8] Figure 14.8 shows typical CBE equipment, almost identical to the MBE equipment except that the materials are supplied by entirely gaseous sources. Molecular beams of group V materials which are cracked in high-temperature cells reach the substrate and begin to grow. The CBE is considered an effective method for GaInAsP systems. An MBE using gas sources only or group V materials is also called a gas source MBE.[9]

14.2. LASER DEVICES AND FABRICATION PROCESSES

14.2.1. Energy Band Structures in Heterojunction Devices

Figure 14.9 shows various possibilities of junctions in semiconductors with different band gaps. We discuss here how the *pn* or *np* junction is formed. The active layer of semiconductor lasers usually has a narrow band

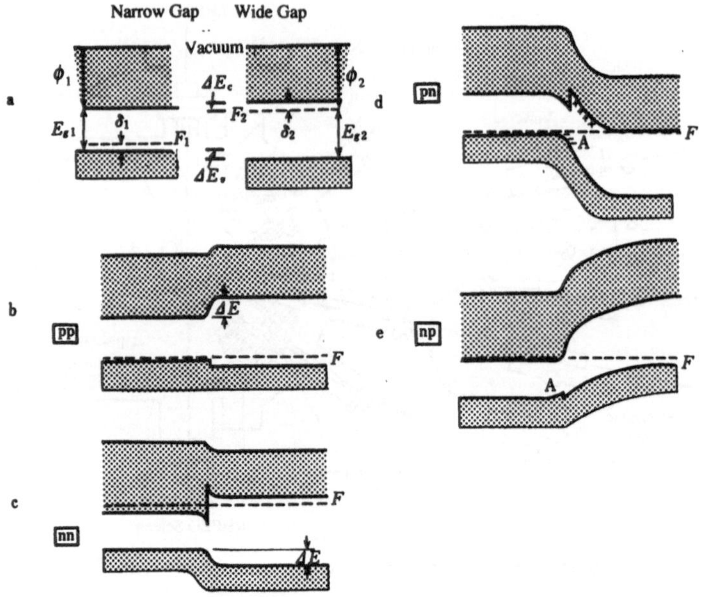

Figure 14.9. Various junctions.

gap, and if the proper two choices of the four in Fig. 14.9b to 14.9e are combined, a double heterostructure is constructed.

We assume that the work functions and the band gaps of two semiconductors, one with a narrow gap and one with a wide gap, are given by ϕ_1 and ϕ_2 and E_{g1} and E_{g2}, respectively (see Fig. 14.9a). When a *pp* junction is formed by doping, the energy levels change as shown in Fig. 14.9b. The Fermi levels on both sides of the junction coincide. There is a slight difference in the valence band because of acceptor levels, and a large difference in the conduction band. The difference in the band gap is largely the result of this change in the conduction band and is expressed as $\Delta E = \Delta E_g = E_{g2} - E_{g1}$. Figure 14.9c shows the band structure where both sides of the junction are *n*-doped. In this case, the band-gap difference is mostly the result of a change in the valence band and is again expressed as $\Delta E = \Delta E_g = E_{g2} - E_{g1}$.

Next, let us consider the case where either a *pn* or an *np* junction is formed, as shown in Fig. 14.9d and 14.9e. Here, the band structure will be detailed by referring to results of analysis by Anderson.

14.2.1.1. *pn* Junction

The locations of the energy bands of semiconductor 1 on the *p* side and semiconductor 2 on the *n* side are expressed by the following equations when the bias V_a is applied to the *pn* junction:

$$\begin{aligned} E_{v1} &= (V_{D1} - V_1)[1 - (1 + x/x_1)^2] && (-x_1 \leqq x \leqq 0) \\ &= V_{D1} - V_1 && (-\infty \leqq x \leqq -x_1) \quad (14.6) \end{aligned}$$

$$E_{c1} = E_{v1} + E_{g1} \quad (14.7)$$

$$\begin{aligned} E_{v2} &= -\Delta E_v - (V_{D2} - V_2)[1 - (1 - x/x_2)^2] && (0 \leqq x \leqq x_2) \\ &= -(\Delta E_v + V_{D2}) + V_2 && (x_2 \leqq x \leqq \infty) \quad (14.8) \end{aligned}$$

$$E_{c2} = E_{v2} + E_{g2} \quad (14.9)$$

where all parameters in the above equations are defined in Table 14.1.

14.2.1.2. *np* Junction

Point A in the *np* junction energy level diagram in Table 14.1 is assumed to be an origin. Again using the parameters from that table, we can write:

Table 14.1. Parameters Used for Calculation of *pn* Junction [After Casey and Panish.]

$\Delta E_c = x_1 - x_2$

$\Delta E_v = (x_1 + E_{g2}) - (x_2 + E_{g1})$

(*pn*) $V_D = E_{g1} + \Delta E_c - \delta_1 - \delta_2$

$V_{D1} = V_D/K \quad V_{D2} = V_D - V_{D1}$

$V_1 = V_a/K \quad V_2 = V_a - V_1$

$$K = 1 + \frac{\varepsilon_1(N_{A1}^- - N_{D1}^+)}{\varepsilon_2(N_{D2}^+ - N_{A2}^-)}$$

(*pn*) $$x_1^2 = \frac{2\varepsilon_1(V_{D1} - V_1)}{q(N_{A1}^- - N_{D1}^+)}$$

$$x_2^2 = \frac{2\varepsilon_2(V_{D2} - V_2)}{q(N_{D2}^+ - N_{A2}^-)}$$

(*np*) $$x_1^2 = \frac{2\varepsilon_1(V_{D1} - V_1)}{q(N_{D1}^+ - N_{A1}^-)}$$

$$x_2^2 = \frac{2\varepsilon_2(V_{D2} - V_2)}{q(N_{A2}^- - N_{D2}^+)}$$

δ_1: Fermi level measured from valence band of semiconductor 1

δ_2: Fermi level measured from conduction band of semiconductor 2

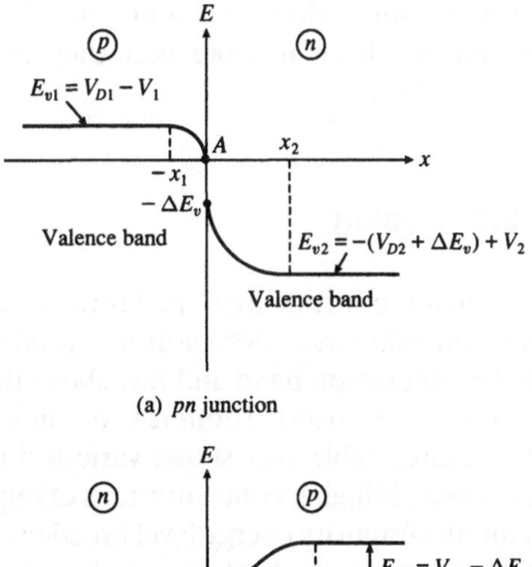

(a) *pn* junction

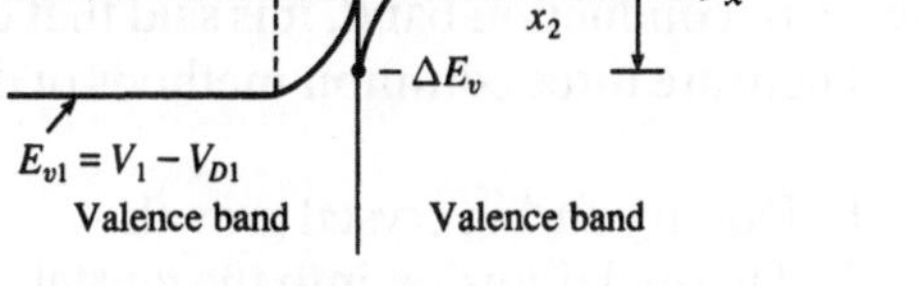

(b) *np* junction

Table 14.2. Impurities Used for Semiconductor Lasers

		n-type	p-type
GaAs/GaAlAs	LPE	Sn, Te, Si, Se	Zn, Ge, Mg, Be
	MOCVD	Se, Si, Ge, Sn	Zn, Mg, Be
	MBE	Sn, Si, Ge	Be, Zn, Mg
InP/GaInAsP	LPE	Sn, Te, Ge, Si	Zn, Cd
	MOCVD	Se, S, Te	Zn
	MBE	Si, Sn	Be

$$E_{v1} = -(V_{D1} - V_1)[1 - (1 + x/x_1)^2] \qquad (-x_1 \leqq x \leqq 0)$$
$$= V_{D1} - V_1 \qquad (-\infty \leqq x \leqq -x_1) \tag{14.10}$$
$$E_{c1} = E_{v1} + E_{g1} \tag{14.11}$$
$$E_{v2} = -\Delta E_v - (V_{D2} - V_2)[1 - (1 - x/x_2)^2] \qquad (0 \leqq x \leqq x_2)$$
$$= -(\Delta E_v + V_{D2}) + V_2 \qquad (x_2 \leqq x \leqq \infty) \tag{14.12}$$
$$E_{c2} = E_{v2} + E_{g2} \tag{14.13}$$

One might think that the *np* junction is just a backward-biased *pn* junction, but we have to note that they are quite different in the case of heterojunctions.

14.2.2. Doping

When impurities are doped into semiconductor crystals, these impurities become donors or acceptors and produce impurity energy levels just below the conduction band and just above the valence band which emit carriers, either electrons (donors) or holes (acceptors) at the ambient temperature. Table 14.2 shows various donor and acceptor materials commonly used. If higher concentrations of impurities are doped into a semiconductor, the impurity energy level broadens to become an energy state (which is called a band tail). In the case where the impurity level is produced in the valence or conduction band, it is said that degeneracy exists.

There are three common methods of doping:

1. Doping during crystal growth
2. Thermal diffusion into the crystal
3. Ion implantation

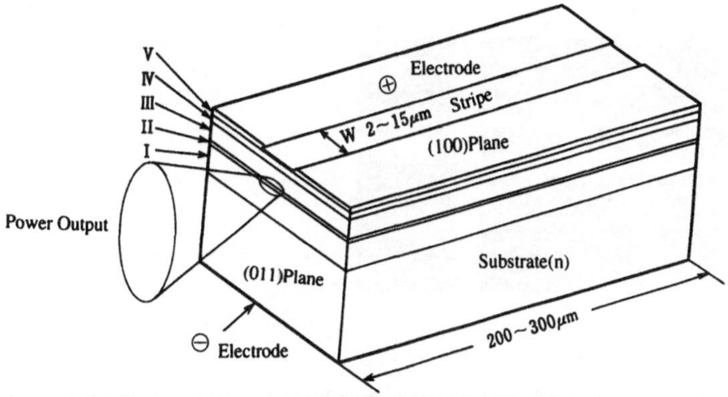

Figure 14.10. Structure of laser chips with double heterostructure.

In conventional stripe lasers, an intentionally undoped active layer is used, but, sometimes, diffusion takes place when zinc is doped into a p-cladding layer in a GaInAsP/InP system. This is called an autodoping method. Especially in a thin-active-layer device ($d = 0.1$ μm or less), the threshold current density can be minimized by choosing an optimum doping concentration. Impurities should be kept low to minimize optical absorption so long as the electrical resistance does not deteriorate.

14.2.3. Fabrication Methods of Wafers for Lasers

14.2.3.1. Structure of Semiconductor Lasers

Figure 14.10 shows the typical structure of a double heterostructure laser device. The location of each layer in the GaAs/GaAlAs and GaInAsP/InP systems is shown in Table 14.3. Layer II is an active layer which is sandwiched between layers I and III which have larger band gaps and smaller refractive indices. As a consequence, charge carriers and light are confined

Table 14.3. Parameters of Double Heterostructure

Layer	Impurity	Thickness (μm)	GaAs/GaAlAs	GaInAsP/InP
IV	p^+	~1	GaAs	GaInAsP
III	p	~3	GaAlAs	InP
II active layer	Nondoped	0.05–0.2	GaAs	GaInAsP
I	n	~3	GaAlAs	InP
Substrate	n	~70	GaAs	InP

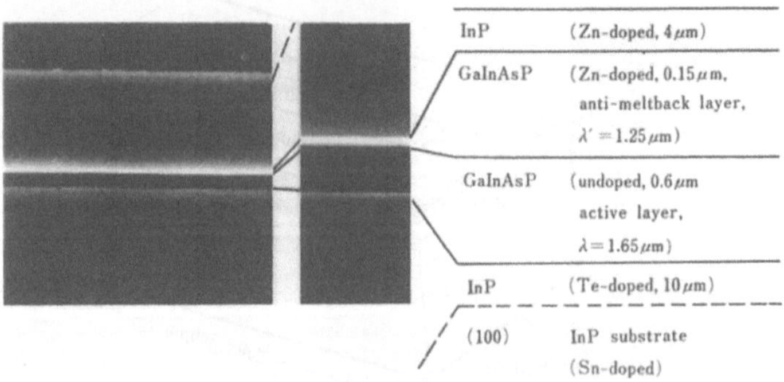

Figure 14.11. Observation of a wafer cross section by SEM. (Photograph courtesy of Professor Y. Suematsu, Tokyo Institute of Technology.)

within the active layer II. A heterojunction device having this type of many-layered structure must be grown maintaining a careful lattice match to the substrate used. In the $GaAs/Ga_{0.7}Al_{0.3}As$ system, a lattice mismatch of +0.04% exists. On the other hand, in the $Ga_xIn_{1-x}As_yP_{1-y}/InP$ ($0 \leq x \leq 0.47$, $0 < y < 1$) system, the lattice mismatch can, in principle, be 0 and is actually $< \pm 0.01\%$ which is the measurement limit.

14.2.3.2. Formation of Multiheterostructure Wafers

Thin-film, multilayered wafers which are lattice matched to a substrate are fabricated by the various epitaxial methods described previously. The LPE method has reliability and is the most common method for production. The MOCVD method is now catching up. In this section the method used to fabricate long-wavelength GaInAsP/InP lasers by LPE is explained as an example. Figure 14.11 shows the cross section of a typical wafer in which the active layer is GaInAsP surrounded by InP cladding layers. To fabricate this device, first n-type InP is grown on an InP substrate. Next, the GaInAsP active layer is formed, then the p-type InP cladding layer is grown. Finally, a p^+-type GaInAsP layer is normally added as a cap layer which has high electrical conductivity. For this device, InP and GaInAsP are alternatively grown on an InP substrate. Molten indium of high purity is used as a solvent and GaAsInAs and InP are used as source crystals of Ga, As, and P. The segregation constant of each of these elements from the liquid phase to the solid phase is different. Over growth temperature ranges from 630 to 650°C, epitaxial growth is possible, maintaining lattice matching with InP $\Delta a/a$

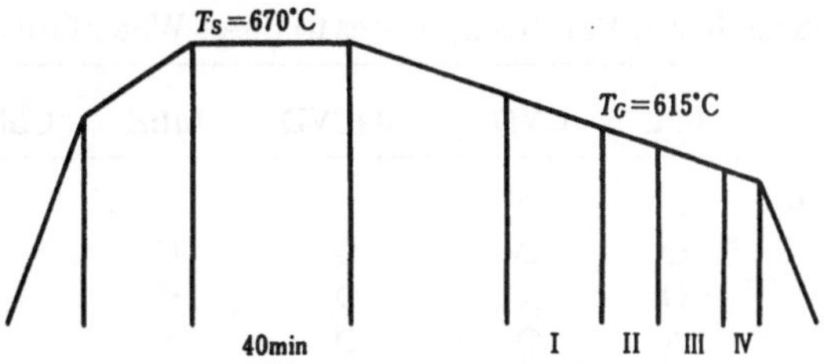

Temperature Program

No	Layer	Content [mg]				
		InP	GaAs	InAs	Te/In	Zn/In
I	InP (Buff)	50			80	
II	GaInAsP (Act)	50	65	225		
III	p-InP (Clad)	50				5
IV	p^+-GaInAsP (Cap)	50	30	155		30
	for 5g of In solvent	Te/In : 2.5% Te in In				
		Zn/In : 1% Zn in In				

Charge

Figure 14.12. Temperature program and charged crystals by a two-phase solution cooling method.

= ±1.01%. Figure 14.12 shows a temperature program and the required crystal charges when a $Ga_{0.2}In_{0.8}As_{0.46}P_{0.54}$ laser with an oscillating wavelength of 1.3 μm is grown by a two-phase solution cooling method.

To observe the active layer, a cleaved cross section is stained with a mixed solution of $K_3Fe(OH)_6$ and KOH so that it can be seen using a scanning electron microscope (SEM). Figure 14.11 shows an SEM micrograph of a typical double heterostructure (DH) wafer. Using this approach to crystal growth of GaInAsP, lasers with outputs ranging from 1.1 to 1.67 μm are available. Other epitaxy techniques are detailed in Table 14.4 which indicates that it is possible to fabricate wafers by VPE, MOCVD, and CBE having thresholds similar to those fabricated by LPE.

14.3. EVALUATION OF WAFERS

This section describes some methods for evaluating grown semiconductor wafers.

Table 14.4. Applicability of Various Epitaxies to Laser Wafer Growth

Crystal system	LPE	CVD	MOCVD	MBE	CBE (gas source MBE)
$Ga_xIn_{1-x}As_yP_{1-y}/InP$	○	○	○	△	○
$Ga_{0.47}In_{0.53}As/InP$	○	○	○	○	○
GaAlAs/GaAs	○	×	○	○	○
GaInAlP	△	○	○	×	○

○, yes.
×, no.
△, difficult.

14.3.1. Observation of Surface Morphology

After the completion of crystal growth, the surface morphology of the wafer is examined either with an optical microscope or with a Nomarski-type differential interference microscope. A mirrorlike surface which cannot be distinguished even with the Nomarski method is essential for laser crystals.

14.3.2. Observation of Cross Sections

Cleaved or diagonally ground sections are stain-etched to emphasize the contrast of the heterostructure layers and are observed with a SEM. Since magnifications up to 200,000 or more are available, an active layer with a film thickness ranging from 20 to 200 Å can be observed. For higher magnifications, a transmission electron microscope (TEM) is employed. Figure 14.13 shows a TEM micrograph of the boundary between InP and GaInAsP.

14.3.3. Determining Composition

The compositions of multilayer samples are determined from the lattice constant by a calibrated x-ray diffractometer, and from the band-gap energy which is measured by photoluminescence spectroscopy.

14.3.4. Determining the Band Gap, E_g

The band gap is determined either by the spectral dependence of the absorption coefficient, α, or by the photoluminescence spectrum.

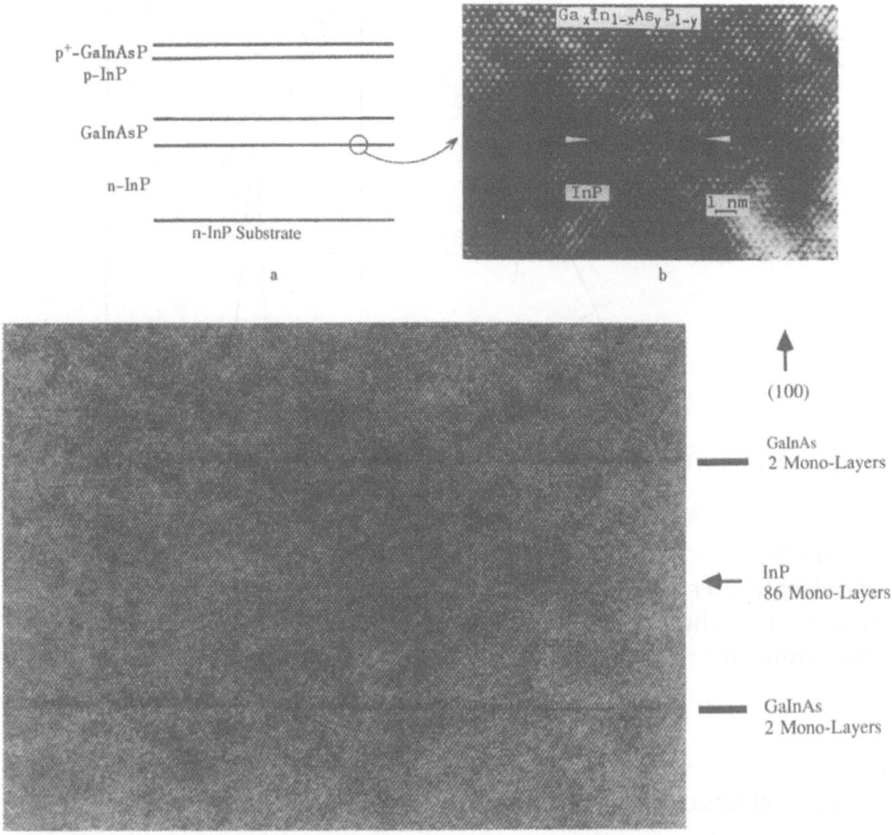

Figure 14.13. (a) GaInAsP/InP lattice images by a TEM. (The wafer was grown in the author's laboratory and observed by Professor Oki of Kyushu University.) (b) TEM image of GaInAs/InP quantum wells grown by chemical beam epitaxy.

14.3.5. Measuring Lattice Matching

Inspection is required for each wafer since the solid-phase component is affected directly by the mole fraction in the melt and the growth temperature. An x-ray diffractometer is commonly used for the measurement of lattice constants. Figure 14.14 shows a sketch of an x-ray diffractometer setup and the associated locking curve.

14.3.6. Photoluminescence

Spectral uniformity of the emission efficiency of wafers, nonradiating dark spots, and nonuniformities in the band gaps, as well as the widths

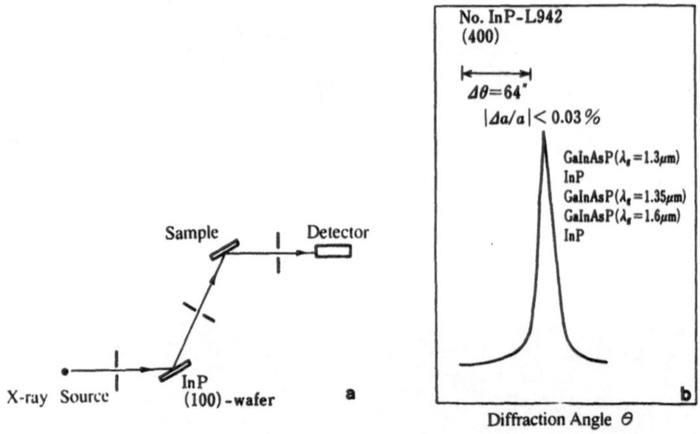

Figure 14.14. (a) Block diagram of dual crystal x-ray diffractometer and (b) locking curve.

of emitting spectra, are measured by exciting the crystal with a short-wavelength laser beam and measuring the photoluminescence. Figure 14.15 shows a photoluminescence setup and a typical photoluminescence spectrum from GaAs.

14.3.7. Measurement of Refractive Index

Since the refractive index differs according to the composition of the heterostructure, values of refractive index versus composition are calculated from the reflection coefficient of light.

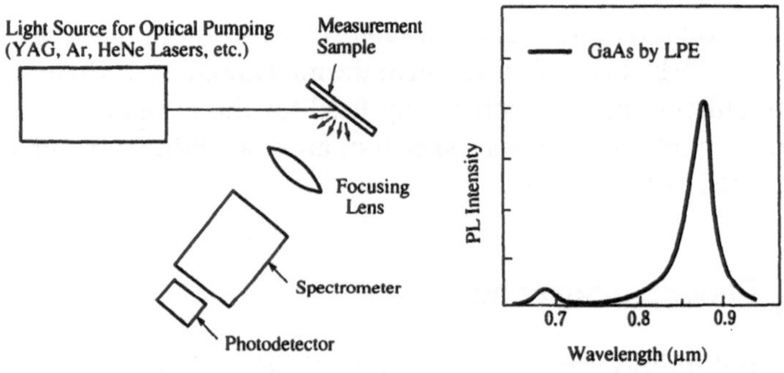

Figure 14.15. Principle of photoluminescence and spectrum from GaAs.

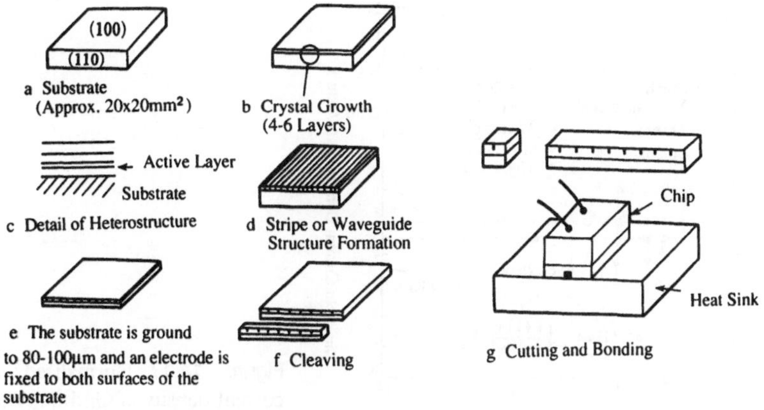

Figure 14.16. Laser fabrication process.

14.3.8. Misfit Density

Etch pit density (EPD) is measured in order to estimate the misfit density of crystals which critically affects laser lifetime. For long-life GaAs/GaAlAs lasers, an EPD of $10^3/cm^2$ or less is required. For the GaInAsP/InP system, and EPD of 10^3 to $10^4/cm^2$ is usable. A substrate with low dislocation densities is, of course, essential for laser reliability.

14.4. FABRICATION OF FUNDAMENTAL LASER DEVICES AND CHARACTERIZATION METHODS

14.4.1. Fabrication Method of Fundamental Laser Devices

The most important step in laser fabrication is the evaluation of their efficiency. This is done after the wafer is processed into a laser. The threshold current density of the laser is measured using wide electrode structures on a device which is called a broad-contact laser. As shown in Fig. 14.10, a broad-contact, 50-μm-wide stripe electrode is placed on the surface of the wafer. After the electrode placement, a laser resonator is fabricated. For example, Fig. 14.10 shows a Fabry–Perot resonator where reflection of light from parallel plane mirrors is utilized. Figure 14.16 summarizes the laser fabrication process, which can, in general, be categorized into the following three processes:

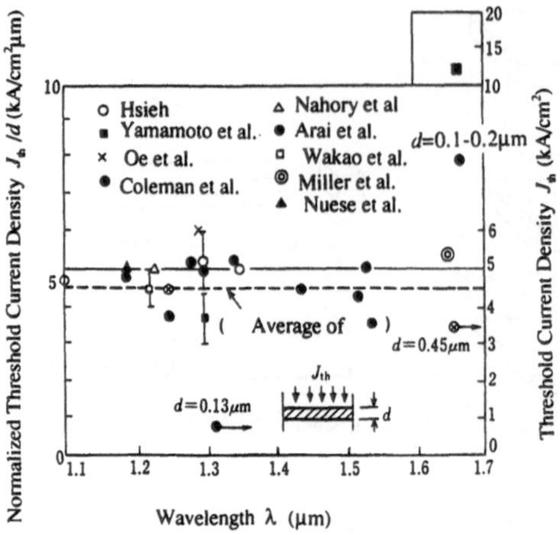

Figure 14.17. Normalized threshold current density of $Ga_xIn_{1-x}As_yP_{1-y}/InP$ laser per unit active layer width.

1. Multilayer wafer growth including the optical waveguide formation process
2. Electrode formation process
3. Resonator fabrication

The wafer fabrication process includes the fabrication of the waveguide to contain the oscillating light.

The laser performance is evaluated by measuring the threshold current density of the laser thus formed. As the current is increased, the output power suddenly increases at a certain level. This level is called the oscillation threshold current, I_{th}. The threshold current density, J_{th}, is then found from the following expression:

$$J_{th} = I_{th}/(LW) \tag{14.14}$$

where L and W are the length of the laser and stripe width, respectively.

J_{th} changes noticeably with the thickness of the active layer, d. The minimum threshold current density occurs around d = 0.1 to 0.2 µm, corresponding to J_{th} = 1 to 2 kA/cm^2 in a typical device. The normalized threshold current density, J_{th}/d, is a standard measure of the quality of the laser crystal. Figure 14.17 shows the threshold current density of $Ga_xIn_{1-x}As_yP_{1-y}/InP$ per unit active layer width.[10] Approximately 4.5 kA/cm^2-µm is typically found in lasers operating between 1.1 and 1.55 µm, which have been fabricated by LPE. Similarly, approximately 5 kA/cm^2-µm is measured in GaInAs/InP structures fabricated by MBE and VPE.

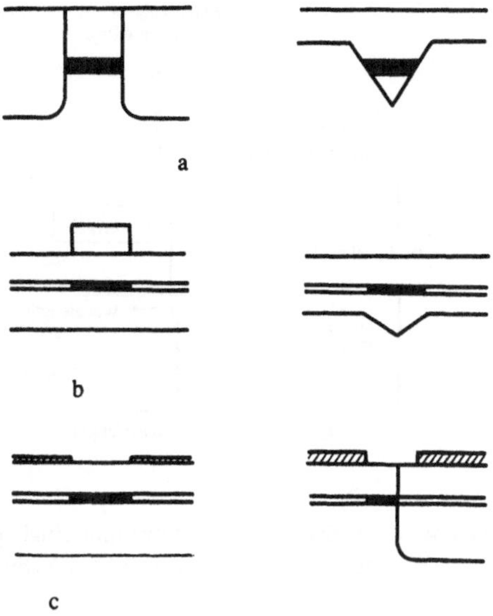

Figure 14.18. Various stripe structures. (a) Buried type; (b) equivalent index waveguide type; (c) impurity concentration difference type. Solid bar shows the active layer.

14.4.2. Stripe-Geometry Lasers

Usually, semiconductor lasers have a stripe electrode structure because the current should be kept small, and for good coherence, oscillation should start as nearly as possible from a point source of light. The structure is required to have small series resistance and excellent thermal dissipation properties. There are various ways of making a striped geometry, and a series resistance of 0.2 to 2 Ω is obtained at an electrode with a width of 1–10 μm and a length of 250 μm. Figure 14.18 shows some stripe electrode structures. In Fig. 14.18a, optical waveguiding occurs in the horizontal direction because of the active layer, so the transverse mode stays stable even when the current is increased. In most cases, a single longitudinal mode can also be achieved. In Fig. 14.18b, optical waveguiding in the horizontal direction is achieved by an equivalent refractive index waveguide. This effect is the result of the difference of refractive indices in the horizontal direction, and additional gain-guiding occurs when the current is increased because of charge carriers. The structure in Fig. 14.18c achieves waveguiding from the difference in impurity concentrations in the horizontal direction.

Typical performance parameters of a GaInAsP/InP stripe laser with an applied voltage of 0.9 V are a current of 10 to 50 mA, a resistance of 1 to 2 Ω, and an output of several milliwatts. It is common for semiconductor lasers to be fabricated with a stripe active layer width between 2 and 15 μm in order to achieve stabilization transverse modes, single longitudinal mode operation, and good linearity in the light output characteristics versus current.

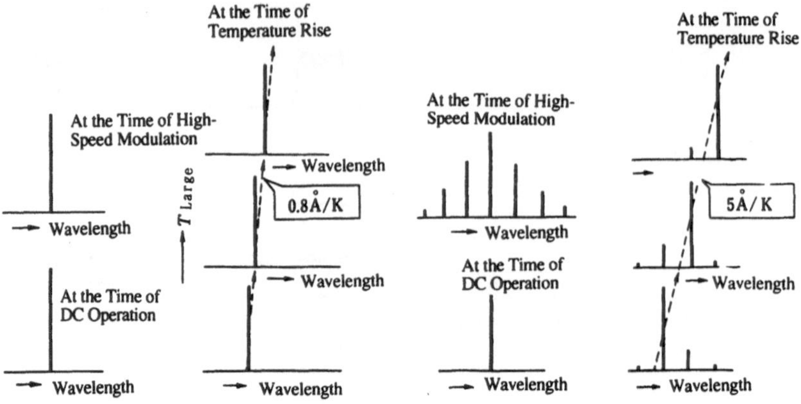

a b

Figure 14.19. Spectrum in (a) dynamic single mode lasers and (b) conventional (long-cavity Fabry–Perot-type) lasers. [After Y. Suematsu, *Proc. IEEE* **71**, 692 (1984).]

Various approaches have been proposed for the fabrication of GaAs/GaAlAs and GaInAsP/InP stripe lasers. A distributed feedback (DFB) laser utilizing a diffraction grating and a distributed Bragg reflector (DBR) laser have been experimentally developed as part of an effort to achieve single longitudinal mode operation. These devices will be detailed later. Furthermore, the temperature characteristics of these devices have been studied. The development of optical integrated circuits employing InP as a substrate is considered to be very important for optical integration techniques in the future. Therefore, chemical etching methods which are capable of discrimination in the (100) and (011) directions, etc. are being studied. Details of this technology will be discussed in the following section.

14.5. LONGITUDINAL MODE CONTROL

In 1978, it was found that a silica fiber exhibits ultralow loss at 1.55 μm. Although scientists recognized the importance of the 1.55-μm band for long-distance transmission, there exists a rather large optical dispersion (1.5 psec/km-Å) at that wavelength, so that a very-short-pulse, high-speed signal cannot be transmitted. The realization of the single-mode laser at 1.55 μm not only enables long-haul and wide-band transmission, but also opens up a wide field of applications for high-performance lasers such as wavelength multiplexing systems.

In a conventional Fabry–Perot-type laser, longitudinal modes increase dramatically in number when the laser is modulated at high frequency, even if that laser can operate continuously at a single mode (see Fig. 14.19). This

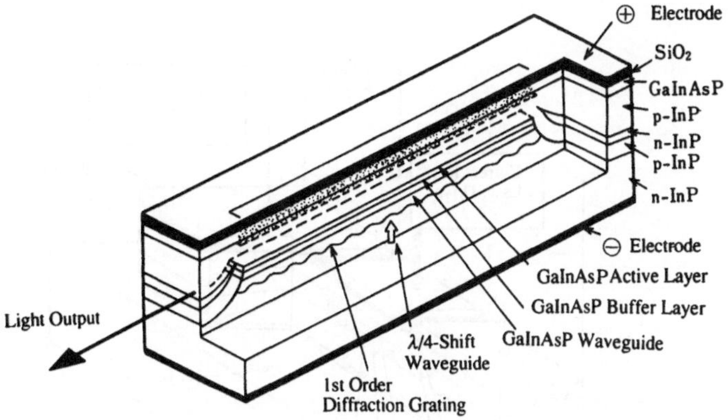

Figure 14.20. Phase shift-type DFB laser. [After K. Utaka, S. Akiba, K. Sakai, and Y. Matsushima, *IEEE J. Quantum Electron.* **QE-22,** 138 (1986).]

can be understood by the fact that, for cw operation, only one mode happens to reach threshold among the very many low-loss modes supported by the cavity because of a small difference in gain. With modulation, however, many modes can reach threshold because of the instantaneous variation of carrier density. It is necessary to arrange for some wavelength selectivity in the cavity configuration so as to maintain single longitudinal mode operation under dynamic conditions (modulation or temperature change). This technology is called longitudinal mode control.

Several schemes exist regarding production of a dynamic single mode laser:

1. Distributed feedback (DFB)
2. Distributed Bragg reflector (DBR) or distributed reflector (DR)
3. Coupled cavity
4. Short cavity
5. Injection locking

We will first consider the DFB laser. Since the resonant frequency is determined by the period of the grating prepared on the active layer, or close to the active layer,[11] single-mode operation can be achieved. Various systems of DFB lasers have been reported. The GaAlAs/GaAs system was the first to be used as a DFB laser, and lasers operating at 1.3 and 1.55 μm are presently in for optical communications. In the DFB laser there are two longitudinal modes near the Bragg wavelength as mentioned in Chapter 8 so one must include an asymmetric structure in the axial direction using a highly reflective mirror, or a phase shifter near the central region (Fig. 14.20) in order to achieve single-mode operation, as was also discussed in Chapter 8.

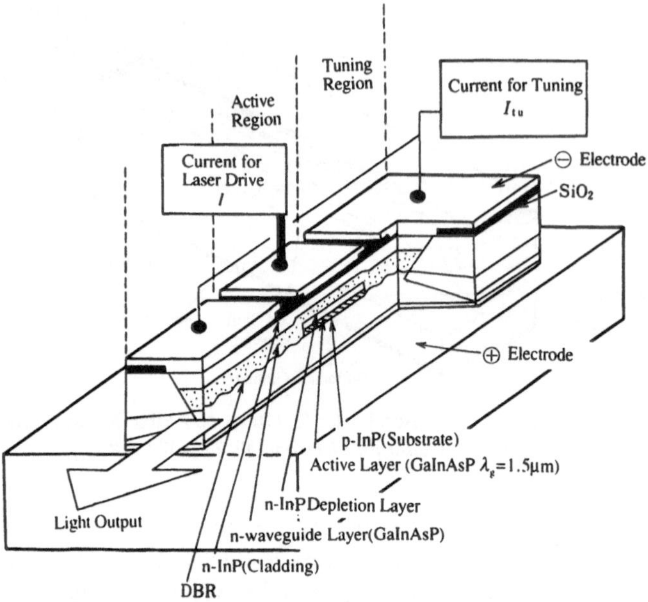

Figure 14.21. Example of a DBR laser. [After Y. Tohmori, H. Oohashi, T. Kato, S. Arai, S. Kimura, H. Hase, and H. Serizawa, *Appl. Phys. Lett.* **45**, 191 (1986).]

In the DBR[12] laser, the grating is prepared on both sides or on one side of the active layer, and acts as a reflecting mirror with frequency selectivity. If the grating is prepared on the same layer as that of the active region, there exists an appreciable amount of absorption so that high reflectivity cannot be achieved. For this reason, a number of methods have been proposed for preparing the grating away from the active layer. An integrated-twin-guide (ITG) structured laser[13] has single-mode operation without mode hopping even when modulated at 1 GHz or higher. As shown in Fig. 14.21, the DBR laser[14] whose output waveguide is directly coupled into inactive waveguide has a high coupling efficiency and fits easily into optical integrated circuits because of the low loss of the waveguide. For this reason, research on DBR lasers has been intensive in the area of photonic integration.

In DFB and DBR lasers the wavelength variation with temperature is as small as approximately 1 Å/K. Again, it is important to stress the fact that one can maintain a single longitudinal mode even when the laser is modulated with a high-frequency signal. This laser is called a dynamic single mode (DSM) laser.[15] It is very important as an optical source in the 1.5 to 1.6 μm wavelength region where ultralow-loss transmission is possible and a narrow linewidth is necessary for wideband communication. The GaInAsP/InP system has been reported to have a lifetime performance better than that of the GaAlAs system.

Table 14.5. Various Types of Dynamic Single-Mode (DSM) Lasers

Integration	Single longitudinal mode operation				
	Distributed Bragg reflector structure	Short resonator structure: Etching	Short resonator structure: Surface emitting	Complex resonator structure	Injection-locking structure
Monolithic formation	O	O	O	O	×
Wafer test	OΔ	O	O	Δ	×
Integration	O	O	O	Δ	×
2-D array	×	Δ	O	×	×
Vertical emission	×Δ	Δ	O	×	×

O, yes.
×, no.
Δ, conditionally yes.

Table 14.5 lists various lasers which can be operated as DSM lasers. DFB lasers operating at 1.3 μm have been in practical use since 1987 for high-speed transmission such as multigigabit per second systems. These lasers in combination with asymmetric phase shift structures and window structures, or DBR-type lasers are crucial for ultrahigh-speed transmission. Furthermore, monolithically fabricated lasers still have future potential, and combining them with an external modulator and amplifiers may become very effective.

14.6. MODULATION AND NOISE

14.6.1. Modulation and Bandwidth of a Semiconductor Laser

In a typical laser the intensity of light is modulated by changing the injection current. When sine-wave modulation, $I(t) = I_m \sin(\omega t)$, is applied around a DC bias current, the intensity of the modulated light can be written as:

$$P(t) = P_0 + P_1 \sin(\omega t + \psi_1) + P_2 \sin(2\omega t + \psi_2) + \cdots \qquad (14.15)$$

Generally speaking, terms higher order than P_2 are small. The modulation coefficient is given as a function of the modulation angular frequency by the solution of a rate equation[16] (see Chapter 12):

$$M = \left| \frac{P_1(\omega)}{P_1(\omega = 0)} \right| = \frac{1}{[\omega^2 \tau_s^2 G + (\omega^2 - \omega_r^2)^2/\omega_r^4]^{1/2}} \qquad (14.16)$$

where f_r is a resonance frequency and is given by:

$$f_r = \frac{\omega_r}{2\pi} = \frac{1}{2\pi\tau_s} \sqrt{\frac{\tau_s}{\tau_p}\left(\frac{I}{I_{th}} - 1\right)} \qquad (14.17)$$

and τ_s denotes the lifetime associated with careful recombination (which is expressed as T_1 in Chapter 12). τ_p is the lifetime of photons within the resonator, and G is a constant. Since $\tau_s = 2$ to 3 nsec, and $\tau_p = 10^{-12}$ sec in the case of a semiconductor laser, the resonance frequency, f_r, is beyond several gigahertz. Modulation is also possible at lower frequencies; however, the characteristics of Eq. (14.17) hold only for homogeneous conditions inside the laser resonator, so at low frequency the performance will change through diffusion of charge carriers and electric impedance.

In the case of high-speed modulation, the wavelength shifts slightly, that is, a chirping of the wavelength occurs which will influence the transmission characteristics through the optical fiber.[17] The amount of chirping is approximately 3 Å at 2 Gbit/sec and is caused by a change in the refractive index related to carrier variation. The design of a structure with small chirping under standard operation conditions is a very important subject.

For the purpose of microwave transmission by the use of lightwave carriers, ultrahigh-speed modulation technology in the range of 30 GHz is developing. Technical issues are aimed at minimizing the capacitance and resistance constants of the device, optimizing laser gain, and finding a new method of modulation. For ultrawideband optical communication, a semiconductor laser which can be stably modulated up to several gigabits per second will be required.

14.6.2. Noise

Intensity fluctuation noise of semiconductor lasers has a peak value at the resonance angular frequency, ω_r. Carrier injection and the reflection of light also cause an increase in the noise characteristics. Therefore, the minimization of light reflection returning to a semiconductor laser is very important. The injection current of the semiconductor laser causes two transient changes: spectral broadening and variation of central frequency. Special care should be taken to minimize these changes for practical use. To minimize spectral broadening, it is necessary to use either a DFB or DBR resonator, or to form a short resonator configuration. The central frequency variation reaches up to 4 Å in a GaInAsP/InP laser. This variation can be used to frequency modulate the laser.

14.6.3. Frequency Stability

Stabilization of the nominal operation frequency, ν, is important for applications requiring interference measurement and for heterodyne communication systems which use the phase of the lightwave. For the measurement of short-term stability or linewidth, spectroscopic analysis is used to observe the noise spectrum around the central frequency, ν_0. In conventional single-mode lasers, a linewidth around several megahertz is considered to be adequate. However, linewidths as low as 7 Hz or less can be obtained by adding frequency control stabilization.

If we express the photon energy, the spontaneous emission coefficient (=2), the photon lifetime in the resonator, and the light power in the resonator as $h\nu_0$, np, τ_p, and P, respectively, then the spectral width, $\delta\nu$, can be given as:

$$\delta\nu = h\nu_0 n_p/(8\pi\tau_p^2 P)(1 + \alpha^2) \tag{14.18}$$

This spectral width, $\delta\nu$, is inversely proportional to the light power and was first derived by Schawlow and Townes. In the semiconductor laser, the term $(1 + \alpha^2)$ must be added, as was pointed out by Henry.[18] α is defined by the ratio of $\delta n'$ to $\delta n''$:

$$\alpha = \delta n'/\delta n'' \tag{14.19}$$

Experimentally observed values of α range between 2 and 7. In addition to the noise described by this expression which is intrinsic to the laser, a much greater noise arises from mode hopping between longitudinal modes in a practical device. In this case, when the ambient temperature changes, or reflected light is coupled back into the cavity, mode hopping occurs in the oscillating longitudinal modes, leading to not only wavelength, but also intensity fluctuations in the output. These fluctuations can be minimized by operating the laser in multimodes, or by superimposing the high-frequency signal onto the DC bias.

The linewidth, $\delta\nu$, must be small (1 MHz or less), especially for coherent optical communication and interference measurements. This is achieved using the following techniques:

1. Stabilization of the center frequency
2. Reduction in linewidth
3. Synchronization with external optical signals (injection locking)
4. Tuning the frequency in reference to a desired value

By integrating these functions, a DSM laser becomes a useful source for coherent optical communications and lightwave sensing.

On the other hand, a high Q-factor is required in order to improve the short-term frequency stability. Furthermore, the influence of reflected light must be eliminated during measurement. The long-term frequency stability is mainly affected by the change of temperature and current noise. As introduced in Chapter 13, the Allan variance[19] $\sigma_y(T)$ expresses the long-term stability with $y = \Delta\nu/\nu$ and T the measurement time period, where the temporal variation of frequency is assumed to be $\Delta\nu$ around the nominal frequency ν. For Allan variance measurement when a semiconductor laser is working continuously and stabilized by utilizing some absorption line center of atoms, the value of $\sigma_y(T) = 10^{-12}$ to 10^{-14} is obtained when $T = 100$ sec, which is comparable to other highly stabilized frequency sources.

14.7. PROSPECTS OF SEMICONDUCTOR LASERS

Predicting the development of semiconductor lasers is very hard, but laser devices must meet the following conditions:

1. Capable of monolithic fabrication
2. Independent process between laser resonator formation and separation of devices
3. Capable of initial tests in wafers
4. Capable of obtaining as large an output as possible from the output facet

Table 14.6 shows expected limits of laser performance. The values in parentheses are typical data that have been obtained to date.

14.7.1. Laser Arrays

In order to obtain a higher output, laser arrays are studied because the practical limit of single stripe lasers is considered to be 100 to 500 mW. In a

Table 14.6. Performances of Semiconductor Lasers[a]

Threshold	<1 mA? (0.9 mA)
Output	>280 mW/device (250 mW)
	>10 W/array (5.4 W)
High-speed modulation	Speed > 30 GHz (30 GHz)
	SI substrate C → small, R → small
	Chirping < 1 Å (3 Å)
Noise (RIN)	RIN < −140 dB/Hz
	Temperature stability
	Feedback noise
	Aiming for multimode → threshold, astigmatism
	Aiming for single mode → single mode techniques
Frequency	Resettability 1 Å
	Stabilization 10^{-14} (10^{-13})
	Linewidth < 1 MHz (~100 kHz)
Transverse mode and coupling efficiency	Wavefront (astigmatism)
	Form: matching to fiber diameter
Temperature characteristics	High temperature of 100 to 250°C

[a] Typical values are given in parentheses.

one-dimensional (1-D) laser array in Fig. 14.22, 3.4 W of continuous output is reported and 2-D laser arrays are also being introduced. Incoherent laser arrays of 10 W or more (each laser has no directivity of independent oscillation or a single peak) are then realized. Actually, 70 W was reported in 1989 and 120 W in 1992. Another important goal is to obtain single-peak coherent laser arrays through phase locking.

14.7.2. Integration

In this section we discuss the research on integration of lasers.

14.7.2.1. Integration of a Laser with Other Photonic Devices

Integration of a laser together with modulators/switches, wavelength tuning elements, amplifiers, and many other photonic devices are considered to be very attractive. Some sophisticated integration has been achieved, i.e., coherent receivers including local oscillators, balanced detectors, and necessary optical waveguides as shown in Fig. 14.23.[23]

However, there is a problem regarding the isolation from reflected light and, therefore, an integrated isolator is required. Along with the develop-

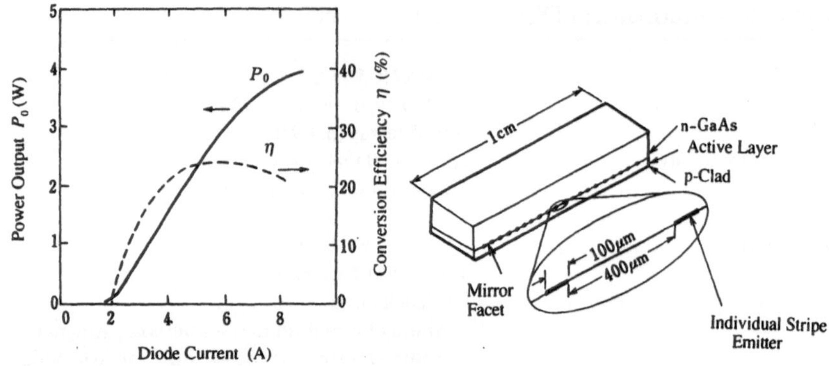

Figure 14.22. One-dimensional laser array. [After D. R. Scifres, C. Lindstrom, R. D. Burnham, W. Streifer, and T. L. Paoli, *Electron. Lett.* **19**, 169 (1983).]

ment in growth technology of magnetic semiconductors by MBE, this problem may be solved.

14.7.2.2. Integration of Peripheral Electronic Circuits

Eventually, the realization of optoelectronic integrated circuits (OEICs) for repeaters where a laser, photodiode, and peripheral electronic circuits are integrated in a monolithic way is a major objective. Figure 14.24 shows an example of an OEIC.[24] The major problem is at which level the OEICs are

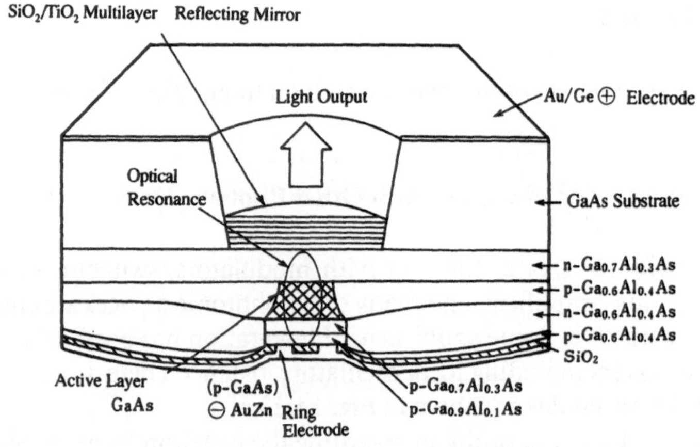

Figure 14.23. A photonic integrated circuit for coherent receivers. (After H. Takeuchi *et al.*[23])

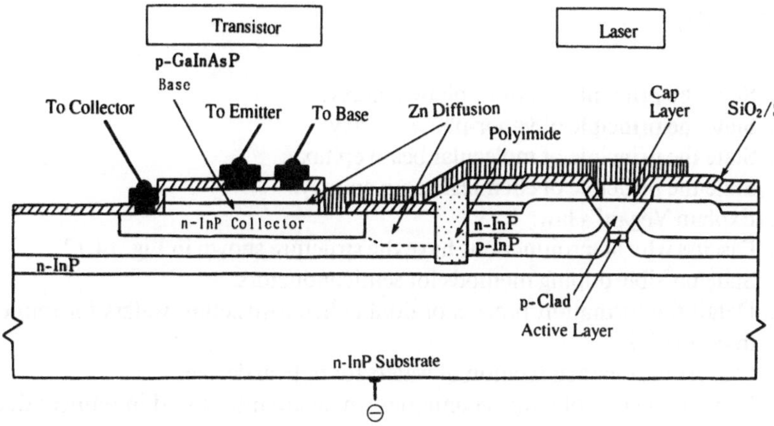

Figure 14.24. Example of an OEIC. (After Shibata *et al.*[24])

put into practical use. When we consider the price, the size of OEICs must be taken into consideration.

14.7.2.3. Integration of Lasers and Photodetectors on Si LSIs

Not only the research on high-speed LSIs which employ compound semiconductors such as GaAs or GaInAs, but also the realization of middle-scale custom LSIs is a very important subject. However, at this stage, the integration method of the minimum numbers of lasers and photodetectors on an Si LSI or a compound LSI is considered to be effective in interconnection of these integrated circuits.

14.7.3. Prospects of Optical Subsystems

Most of the lasers are being utilized in various research and industrial areas. I hope that lasers can generate larger-scale industries. For this progress, we must improve not only laser performance, but also many associated optoelectronic components. For example, light should be guided through an optical fiber by simply coupling with an optical connector without radiating it into the air, and integrated optics may realize optical devices integrated into a one-chip circuit. We hope to replace a gigantic optical bench and handmade subsystems.

PROBLEMS

14.1. State the principle of liquid-phase epitaxy.
14.2. State the principle of vapor-phase epitaxy.
14.3. State the principle of molecular beam epitaxy.
14.4. State the principle of chemical beam epitaxy.
14.5. Explain Vegard's law.
14.6. Discuss why heterojunctions have the structure shown in Fig. 14.12.
14.7. State possible doping methods for semiconductors.
14.8. Detail the formation process of double heterostructure wafers for semiconductor lasers.
14.9. Discuss the characterization of wafers for optical devices.
14.10. Explain the role of stripe geometries which are introduced in semiconductor lasers.
14.11. Consider why a gain-guided laser can operate in multilongitudinal modes.
14.12. State the necessity and advantage of a dynamic single mode laser.
14.13. What determines the quantum efficiency of lasers?
14.14. What are the astigmatism of semiconductor lasers and its origins?
14.15. Consider the resonant frequency of a DBR laser. Hint: Use the result of Chapter 8 and the phase shift given by Eq. (8.101) or (8.103).
14.16. Obtain the modulation efficiency given by Eq. (14.16). Obtain higher-order harmonics such as P_2, P_3, etc.
14.17. Consider the physical meaning of the linewidth enhancement factor α.
14.18. What are important features of laser arrays?
14.19. Discuss the importance of integration technology in optoelectronic devices.

REFERENCES

1. K. Iga and K. Moriki, *Handbook on Compound Semiconductor Devices,* Science Forum (1968).
2. C. J. Neuse and G. H. Olsen, *Appl. Phys. Lett.* **26,** 528 (1975).
3. T. Mizutani, M. Yoshida, A. Usui, H. Watanabe, T. Yuasa, and I. Hayashi, *Jpn. J. Appl. Phys.* **19,** L113 (Feb. 1980).
4. M. Razeghi, J. C. Bouley, K. Kazmierski, M. Papuchon, B. de Cremoux, and J. P. Duchemin, *9th IEEE Semiconductor Laser Conf.,* A-4 (1984).
5. R. D. Dupuis and P. D. Dapkus, *Appl. Phys. Lett.* **31,** 839 (1977).
6. W. T. Tsang, *IEEE J. Quantum Electron.* **QE-20,** 1119 (1984).
7. B. I. Miller, J. H. McFee, R. J. Martin, and P. K. Tien, *Appl. Phys. Lett.* **33,** 44 (1978).
8. W. T. Tsang, *Appl. Phys. Lett.* **45,** 1234 (1984).
9. H. Temkin, M. B. Panish, R. A. Logan, and J. P. van der Ziel, *9th IEEE Semiconductor Laser Conf.,* F-1, 76 (1984).
10. E. C. Casey Jr. and M. B. Panish, Heterostructure Lasers, Part A & B, Academic Press, New York (1978).
11. H. Kogelnik and C. Shank, *J. Appl. Phys.* **43,** 2327 (May 1972).
12. W. Tsang and S. Wang, *9th IQEC,* 38 (June 1976).

13. K. Utaka, K. Kobayashi, F. Koyama, Y. Abe, and Y. Suematsu, *Electron. Lett.* **17,** 368 (May 1981).
14. Y. Abe, K. Kishino, Y. Suematsu, and S. Arai, *Electron. Lett.* **17,** 945 (Dec. 1981).
15. K. Iga and Y. Suematsu, *1st Eur. Conf. Integrated Opt.*, 70 (Sept. 1981); Y. Suematsu, *Proc. IEEE* **71,** 692 (1984).
16. T. Ikegami and Y. Suematsu, *Trans. IEICE Jpn.* **51-B**(2), 57 (1967).
17. F. Koyama, S. Arai, Y. Suematsu, and K. Kishino, *Electron. Lett.* **17,** 983 (Dec. 1981).
18. C. H. Henry, *IEEE J. Quantum Electron.* **QE-18,** 259 (1982).
19. D. Allan, *Proc. IEEE* **54,** 221 (1966).
20. H. Soda, Y. Motegi, and K. Iga, *IEEE J. Quantum Electron.* **QE-19,** 1035 (1983).
21. K. Utaka, S. Akiba, K. Sakai, and Y. Matsushima, *IEEE J. Quantum Electron.* **QE-22,** 138 (1986).
22. Y. Tohmori, H. Oohashi, T. Kato, S. Arai, S. Kimura, H. Hase, and H. Serizawa, *Appl. Phys. Lett.* **45,** 191 (1986).
23. H. Takeuchi, K. Kasaya, Y. Kondo, H. Yasaka, K. Oe, and Y. Imamura, *7th Int. Conf. Integrated Opt. Opt. Fiber Commun.* (IOOC '89), Kobe (Japan), 20PDB-6, Digest No. 5, p. 48 (July 1989).
24. J. Shibata, I. Nakao, Y. Sasai, S. Kimura, N. Hase, and H. Serizawa, *Appl. Phys. Lett.* **45,** 191 (1984).
25. D. R. Scifres, C. Lindstrom, R. D. Burnham, W. Streifer, and T. L. Paoli, *Electron. Lett.* **19,** 169 (1983).
26. D. R. Scifres, *CLEO '86,* TuB1 (1986).
27. D. W. Nam, R. G. Waarts, D. Welch, and D. R. Scifres, *CLEO '92,* CWN8 (1992).

FIFTEEN

SURFACE-EMITTING LASERS

In this last chapter, we will consider surface-emitting (SE) lasers. They constitute a new type of semiconductor laser, of which output is taken vertically from the substrate. There are four types of SE lasers: (1) vertical cavity type, (2) grating coupled type, (3) 45° deflecting mirror type, and (4) folded cavity type. In particular, we will discuss the important factors in vertical cavity SE lasers. Some lasing characteristics, device design, state-of-the-art performances, ultimate performances, and possible future applications will also be discussed.

15.1. ADVANTAGES OF SURFACE-EMITTING LASERS

Large-scale parallel lightwave communication systems, quick-access optical disks, optical computing, optical interconnects, and so on are accelerating the necessity of the SE laser. The author suggested a vertical cavity SE laser in 1977.[1] Figure 15.1 shows a model of a Fabry–Perot resonator in a vertical cavity SE laser, normally employing a radial symmetry. The cavity is formed by two surfaces of an epitaxial layer, and light output is vertically taken from one of the mirror surfaces. Other types of SE lasers employing horizontal in-plane cavity structures have also been studied.

Figure 15.2 shows the first SE laser, fabricated in the author's laboratory.[1] It has been put to practical use through studies since 1978.[2] The SE laser has various features such as vertical emission, single mode, and possibility of 2-D array as shown in Table 15.1.

Several types of SE lasers have been developed, the light output of which

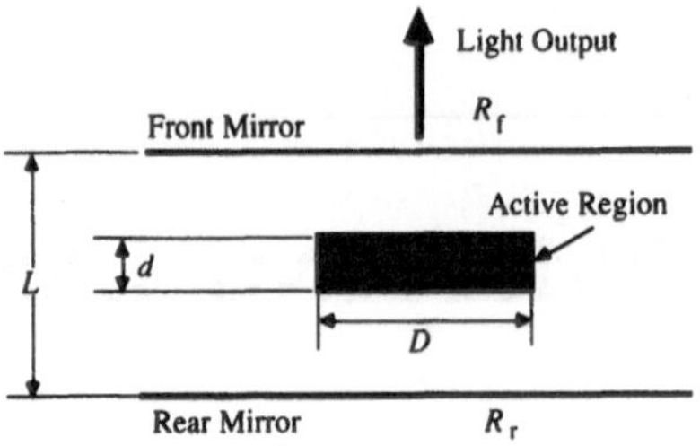

Figure 15.1. Typical model of a surface-emitting laser.

is taken vertically from the substrate. They can be classified into the following types:

1. Vertical cavity SE laser: A laser whose microstructure is prepared with a reflecting mirror on both sides of an epitaxial layer.
2. Grating coupled SE laser: A laser which can output light perpendicularly by a second-order diffraction grating.
3. 45° deflecting mirror SE laser: A stripe laser to which an external 45° reflecting mirror is attached to deflect light.
4. Folded cavity SE laser: A laser having an intracavity 45° mirror to fold a cavity.

The last three types of lasers have been developed for the purpose of obtaining very high powers, such as more than several or several tens of watts in continuous wave (cw) operation. The vertical cavity device may find a good application by virtue of forming a densely packed 2-D array.

SE lasers have a number of advantages:

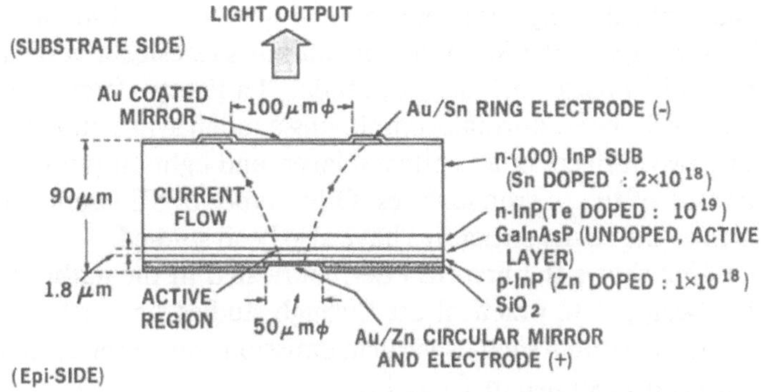

Figure 15.2. The first surface-emitting laser, fabricated in 1979.

Table 15.1. Features of Surface-Emitting Lasers

	Laser characteristics	2-D Laser array capability	Coupling with other devices
Vertical cavity 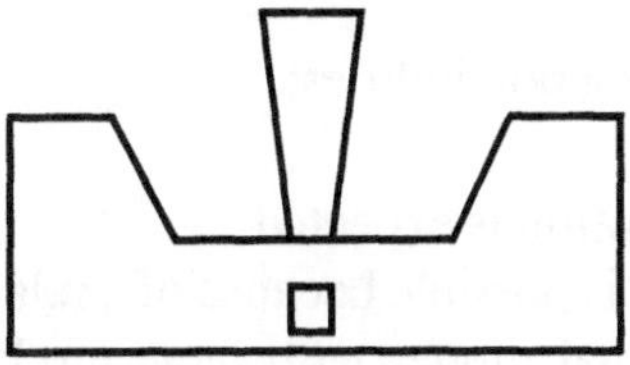	Narrow circular beam Single-mode operation	Free arrangement Dense packing	Vertical stacking
Grating coupled 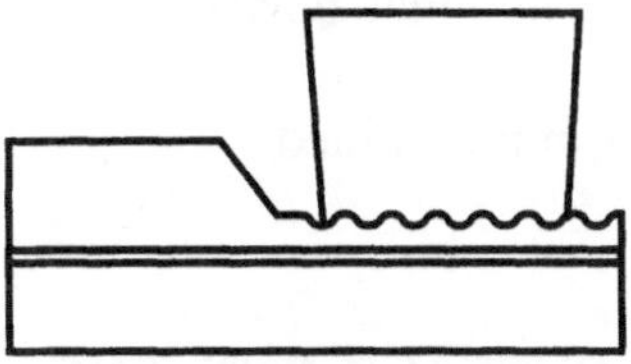	Narrow beam in one direction Single-mode operation	Limited by cavity length 2-D phase-locking	Beam angle sensitive to the change of wavelength
45° Mirror 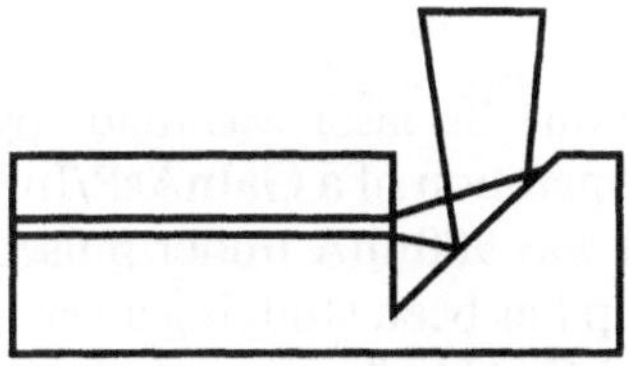 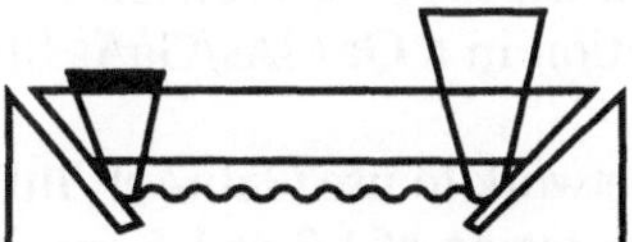	Compatible with conventional structure Beam quality dependent on mirror flatness	Limited by cavity length	Similar to stripe lasers
Folded cavity 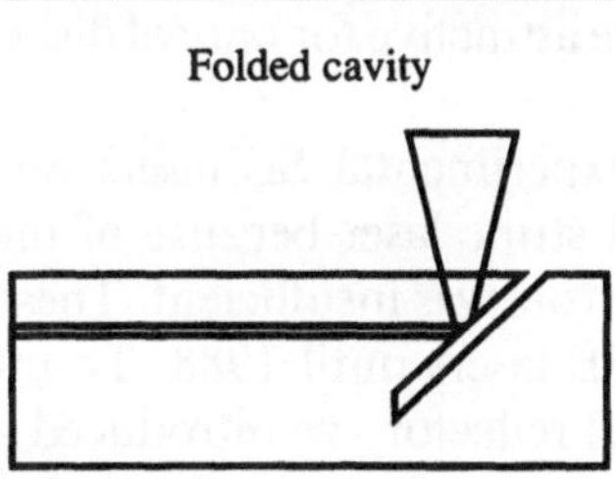	Limited equivalent reflectivity	Limited by cavity length	Similar to stripe lasers
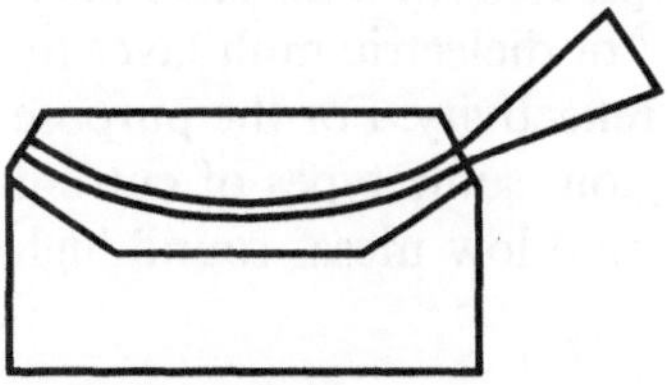	Simple to manufacture	Difficult because of oblique output beam	Similar to stripe lasers

1. A large number of laser chips can be fully monolithically processed.
2. The probe test can be performed before separation into chips.
3. The laser beam can be radiated out perpendicular to the wafer.
4. A large-scale 2-D laser array can be formed.
5. High output power is possible with 2-D arrays.

The vertical cavity device offers the following possibilities:

6. Operation with ultralow power consumption is expected.
7. Dynamic single mode (DSM) operation is possible because of a relatively large gain difference in longitudinal modes with large mode spacing (=100–400 Å).
8. Stack integration with multi-thin-film functional devices can be made intact to an SE laser.
9. A circular beam with negligible astigmatism is obtained.

15.2. HISTORY OF VERTICAL CAVITY SURFACE-EMITTING LASERS

The first suggestion of a double-heterostructure SE laser was made by the author in 1977. We obtained the first lasing operation of a GaInAsP/InP SE laser device in 1979, in which the threshold was 900 mA under pulsed operation at 77 K.[1] Since then, the author's group has been studying a vertical cavity SE laser device with GaInAsP/InP and GaAlAs/GaAs systems.[2–6] We obtained a room-temperature pulsed operation in a GaAlAs/GaAs SE laser in 1985.[5]

It is advisable for a long-haul system and network to use GaInAsP/InP SE single-mode lasers emitting in the wavelength region of 1.3 or 1.5 μm, if realized. On the other hand, GaAlAs SE lasers are attractive for optical disks, optical sensing, and optical parallel processing.

The threshold current density of initial experimental SE lasers was rather high in comparison with a conventional stripe laser because of the short gain region, when the reflectivity of the mirrors was insufficient. These prevented room-temperature cw operation of SE lasers until 1988. To increase the reflectivity of the *p*-side (bonding side) reflector, we introduced a ring electrode in which the reflecting mirror is separated from the electrode.[6] In addition, we introduced an Au/SiO_2 mirror,[7] or dielectric multilayer reflector,[8] for improving the *n*-side (output side) reflectivity. For the purpose of effectively confining current in an active region, some types of current confining structures were introduced, i.e., a round-low mesa, round-high

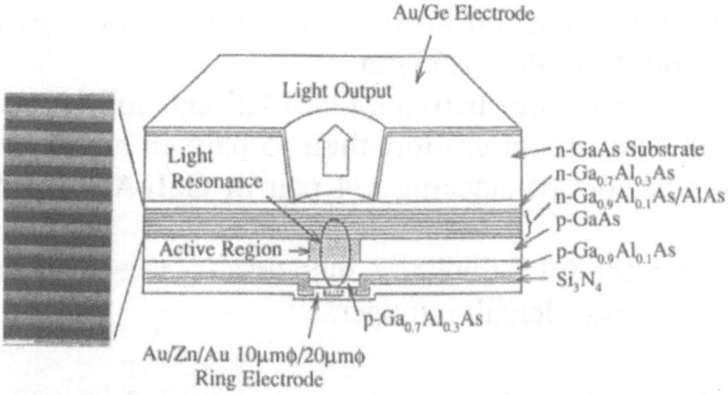

Figure 15.3. A cross-sectional scanning electron microscope (SEM) photograph of a wafer with $Ga_{0.9}Al_{0.1}As/AlAs$ multilayer Bragg reflector and an applied surface-emitting (SE) laser. [After Sakaguchi *et al.*[15]]

mesa/polyimide buried, and circular buried heterostructure (CBH).[9] By introducing a CBH, the threshold was dramatically reduced and low threshold was obtained in a GaAlAs/GaAs system.[10,11] In 1988, we achieved the first room-temperature cw operation of a GaAlAs/GaAs SE laser.[12,13] After we demonstrated some good characteristics of cw SE lasers, much attention was paid to SE lasers and many research groups began research on the vertical cavity SE laser.

Vertical cavity SE lasers utilizing semiconductor multilayer reflectors as shown in Fig. 15.3 for a DBR[14,15] or DFB[16,17] laser may enable the integration of thin-film functional optical devices onto an SE laser by stacking them. This will open up a new scheme for 3-D integrated optics.

15.3. VERTICAL CAVITY SURFACE-EMITTING LASERS (VCSEL)

15.3.1. GaInAsP/InP Surface-Emitting Lasers

The importance of SE lasers emitting at 1.3 or 1.55 μm is currently increasing, since their applications to parallel lightwave systems and parallel optical interconnects are actually being considered. Although the GaInAsP/InP VCSE laser is expected to be realized, this material system has substantial difficulties regarding fabrication of the VCSE laser:

1. The Auger nonradiative recombination is noticeable and this makes the threshold carrier density abnormally high near room temperatures.

2. There is a resonant intra-valence-band absorption, particularly for the compositions near 1.55 μm.
3. The index difference between GaInAsP and InP is relatively small ($\cong$10%), which requires more than 35 pairs in preparing distributed Bragg reflectors by stacking $\lambda/4$ pair of GaInAsP/InP heterostructure.
4. Moreover, the total thickness increases to several microns, simply because the wavelength is longer.

For these reasons, we have not as yet succeeded in achieving room-temperature cw operation in this material. The key issue in terms of this material for VCSE lasers is, therefore, to have a highly reflective mirror and build an effective current injection scheme. For this purpose, a GaInAsP/InP SE laser with a CBH[18] modified by using a flat-surface CBH as shown in Fig. 15.4 was fabricated.[19] This laser can be grown by three-step epitaxies including regrowth processes followed by a successive fully monolithic fabrication process. The substrate is polished to 150 μm thick and the *n*-side Au/Sn electrode is formed. Next, the substrate and etch stop layer are preferentially etched off to form a short cavity (=7μm). One of the favorable characteristics of this material system is that because the chemical etch rate difference for GaInAsP and InP is relatively large, preferential etch can be completely successful. Finally, SiO_2/Si reflectors are formed both on the bottom of an etched well and on the *p*-side epitaxial surface by an electron beam evaporation employing optical thickness monitoring.[20,21]

A current–light output (*I–L*) characteristic of a flat-surface CBH VCSE laser device at 77 K under cw condition is measured. The minimum cw threshold range is 4.6 mA for an 18-μm-diameter active region. The threshold current density is around 2 kA/cm^2. (1 kA/cm^2 corresponds to 10 μA/μm^2.) Pulsed operation has been obtained at near room temperature[19,20] and room temperature.[22] A polyimide buried structure device is fabricated

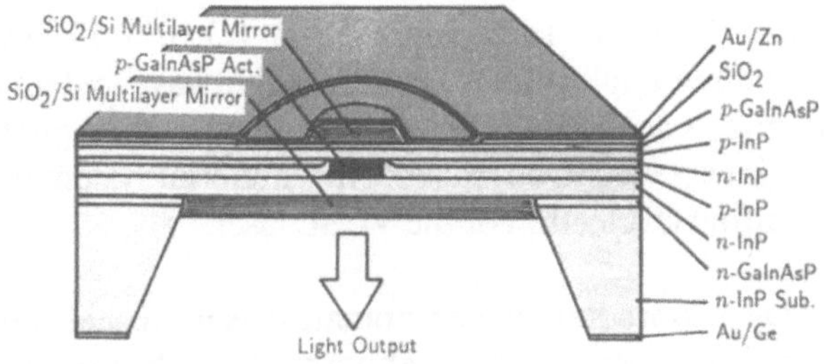

Figure 15.4. A flat-surface circular buried heterostructure surface-emitting (SE) laser. [After Baba *et al.*[19]]

by much simpler processes[9] and pulsed operation at 66°C has been obtained.[23] However, it is still difficult to operate cw at room temperature in this system.

15.3.2. GaAlAs/GaAs Surface-Emitting Lasers

A GaAlAs/GaAs laser can employ almost the same CBH structure as the GaInAsP/InP laser. In order to decrease the threshold, the active region is also constricted by the selective meltback method. In 1987, a threshold of 6 mA was demonstrated for an active region diameter ~6 μm under pulsed operation at 20.5°C.[10] The threshold current density is about 20 kA/cm^2. It is to be noted that a microcavity 7 μm long and 6 μm wide was realized. At present, SE lasers exhibiting $I_{th} \cong 2$–10 mA and a few milliwatts of output power are available at the laboratory level.

The MOCVD-grown CBH SE laser of Fig. 15.5[12,13] was demonstrated by a two-step MOCVD growth and fully monolithic technology. First, a GaAs/GaAlAs double heterostructure (DH) wafer with an active layer thickness of 3 μm was grown by MOCVD at 780°C under atmospheric pressure.

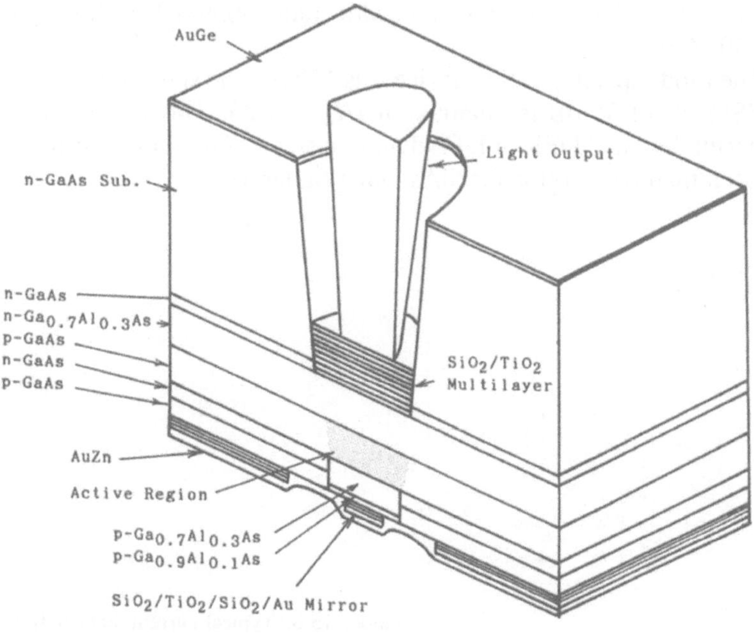

Figure 15.5. A metal–organic chemical vapor deposition (MOCVD)-grown GaAlAs/GaAs surface-emitting laser exhibiting the first room-temperature cw operation. [After Koyama *et al.*[12]]

After the first growth, a silicon nitride (Si_3N_4) circular mask with a diameter of 10 μm was formed on the wafer for mesa etch and selective regrowth of current-blocking layers. A *p*-cladding GaAlAs layer is lightly etched by a sulfuric acid:water:hydrogen peroxide (1:8:8) solution. Selective MOCVD regrowth of GaAs under atmospheric pressure is employed to form current-blocking layers (0.7-μm-thick *n*-GaAs and 0.3-μm-thick *p*-GaAs). The growth condition is the same as that used for the DH wafer growth. There is no deposition on the top of a circular mesa covered with a Si_3N_4 mask. A short cavity structure with a cavity length of 5 μm is formed by selectively removing the GaAs substrate. A ring electrode with outer/inner diameter of 40 μm/10 μm was adopted and SiO_2/TiO_2 mirrors are prepared to both surfaces.

Figure 15.6 shows a typical current–light output characteristic and its lasing spectrum under cw operation at 20°C. The lowest cw threshold current is 30 mA ($J_{th} \cong 26$ kA/cm^2). The differential quantum efficiency is typically 10%. The maximum cw output power is about 2 mW. Saturation of the output power results from a temperature increase of the device. Stable single-mode operation is not observed with subtransverse modes or with other longitudinal modes as shown in the inset of Fig. 15.6. The spectral linewidth above the threshold is less than 1 Å which is limited by the resolution of the spectrometer (ANRITSU MS9001A). We performed a high-resolution measurement of its spectral linewidth and observed a linewidth of about 50 MHz.[24]

The mode spacing of this device was 170 Å. The submode suppression ratio (SMSR) of 35 dB is obtained at $I/I_{th} = 1.25$. This is comparable to that of well-designed DBR or DFB dynamic-single-mode lasers. Many efforts have been made to develop GaAlAs/GaAs SE lasers.[25–33]

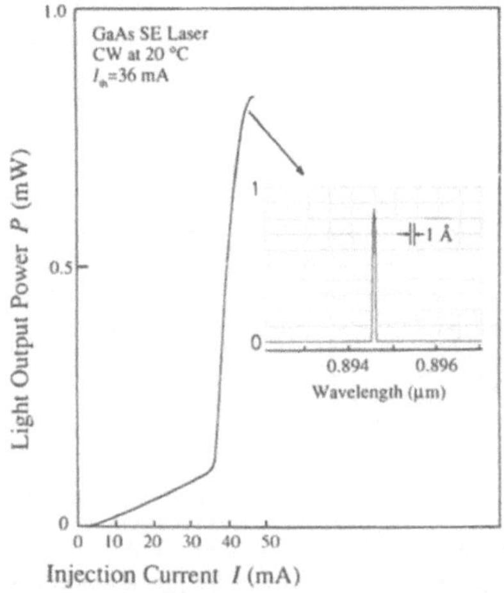

Figure 15.6. Typical current–light output characteristic and lasing spectrum of GaAlAs/GaAs surface-emitting laser under cw operation at room temperature. [After Koyama *et al.*[12]]

15.3.3. GaInAs/GaAs Surface-Emitting Lasers

The GaInAs/GaAs strained pseudomorphic system grown on a GaAs substrate has been developed as a high-power laser emitting at 0.98 μm for pumping an erbium-doped fiber amplifier. This material exhibits a high laser gain and has been introduced into SE lasers together with GaAs/AlAs multilayer reflectors. A low threshold ($\cong$1mA) has been demonstrated by Jewell *et al.* from the microdevices as shown in Fig. 15.7.[33] The threshold current density is about 8 kA/cm^2. The single or multiquantum wells are employed in the active region. The use of thin active layers which are much smaller than the wavelength can enhance the optical confinement factor by 2 if they are positioned properly, i.e., at the node of standing waves. The minimum threshold reported so far is 0.7 mA using this system and the laser shown in Fig. 15.8.[34] The minimum J_{th} is 800 A/cm^2, which is approaching the level of conventional stripe lasers.

15.4. ULTIMATE THRESHOLD AND SPONTANEOUS EMISSION CONTROL

15.4.1. Ultimate Threshold

It is expected that we can obtain a 1-μA device[35] if we can overcome some technical problems, such as making tiny structures, the ohmic resistance of electrodes, and improving heat sinking. Many efforts toward improving the characteristics of SE lasers have been made as shown in Fig. 15.9 including surface passivation in the regrowth process for buried heterostructure, microfabrication, and fine epitaxies.

15.4.2. Spontaneous Emission Control

The spontaneous emission has been considered not to be subject to artificial control. But it will be possible if we can make a special laser cavity, e.g., closed cavity[36] or very tiny resonator as small as the wavelength.[37] When the SE lasers were introduced, spontaneous emission control was considered to occur by taking advantage of the microcavity structure. The spontaneous emission factor has been estimated on the basis of 3-D mode density analysis.[38,39] The possibility of devices with no distinct threshold is suggested.[37]

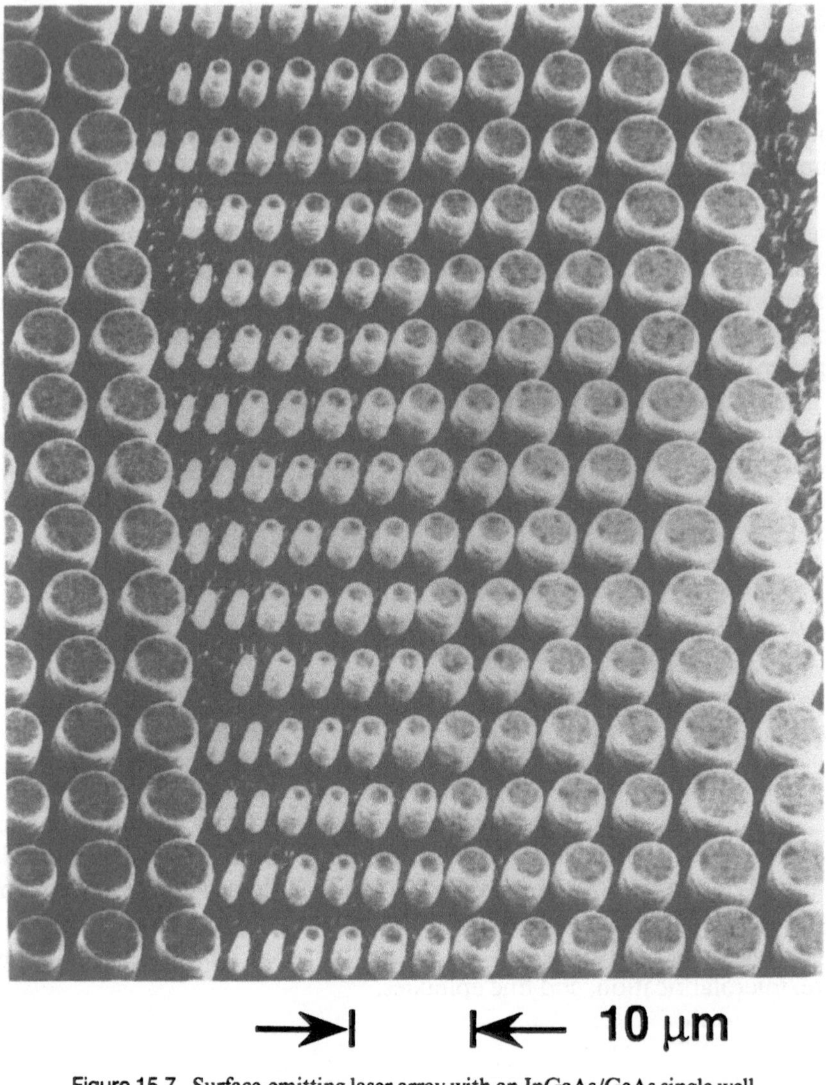

Figure 15.7. Surface-emitting laser array with an InGaAs/GaAs single well. [After Jewell *et al.*[33]]

Figure 15.10 shows the spontaneous emission factor of microcavity VCSE lasers.[38]

15.4.3. Photon Recycling

Microcavity SE lasers are useful for realizing so-called photon recycling, i.e., by covering the side-bounding surfaces of the cavity, some amount of

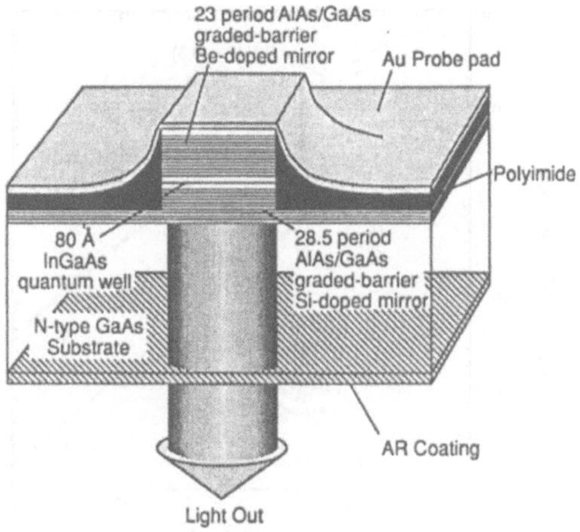

Figure 15.8. Microstructure surface-emitting laser polyimide-buried structure. [After Geels and Coldren[34]]

spontaneously wasted photons can be recycled. It has been demonstrated that by this method, the device appeared to have no distinct threshold as shown in Fig. 15.11.[40] Tamanuki *et al.* showed that the rate of photon recycling could be about 30%.[41]

15.5. TWO-DIMENSIONAL ARRAYS OF SURFACE-EMITTING LASERS

A 2-D array has also been demonstrated for the purpose of constructing high-power lasers and coherent arrays.[42] The coherent coupling of these ar-

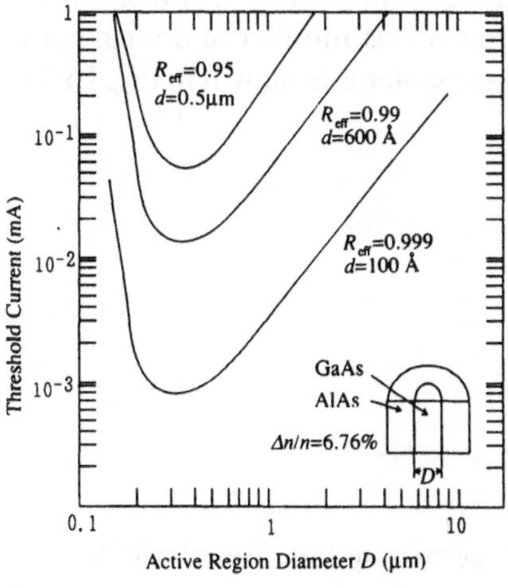

Figure 15.9. Threshold versus active region diameter for GaAs surface-emitting lasers. [After Tamanuki *et al.*[35]]

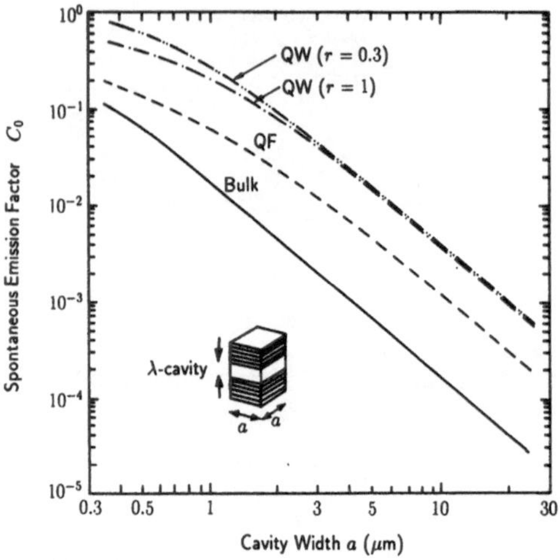

Figure 15.10. Estimated spontaneous emission factor. [After Baba *et al.*[38]]

rayed lasers has been attempted by using a Talbot cavity[43] and diffraction coupling. It is pointed out that 2-D arrays are more suitable for making a coherent array than a linear configuration, since we can take the advantage of 2-D symmetry.[44] The multiwavelength emission from 2-D arrayed SELs was demonstrated as shown in Fig. 15.12 by Chang-Hasnain *et al.* in 1991.[45] This type of device should be used in wavelength domain multiplexing (WDM) lightwave systems.

15.6. APPLIED SUBSYSTEMS

By taking advantage of the 2-D arrayed configuration available in SE lasers, we can expect to simultaneously align a vast number of optical components, as in parallel multiplexing lightwave systems as shown in Fig. 15.13.

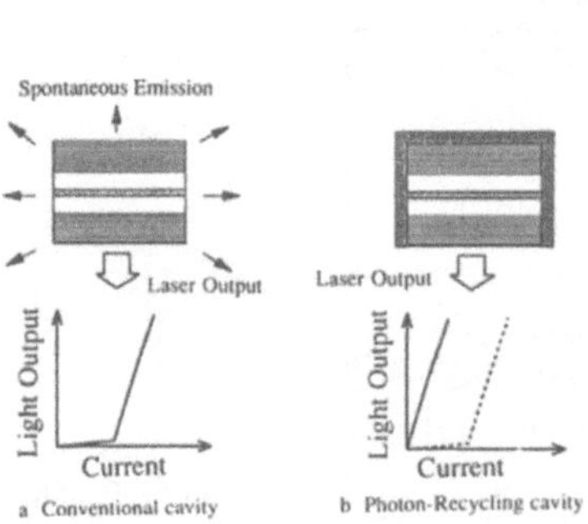

Figure 15.11. A VCSE laser using a photon-recycling structure. [After Numai *et al.*[40]]

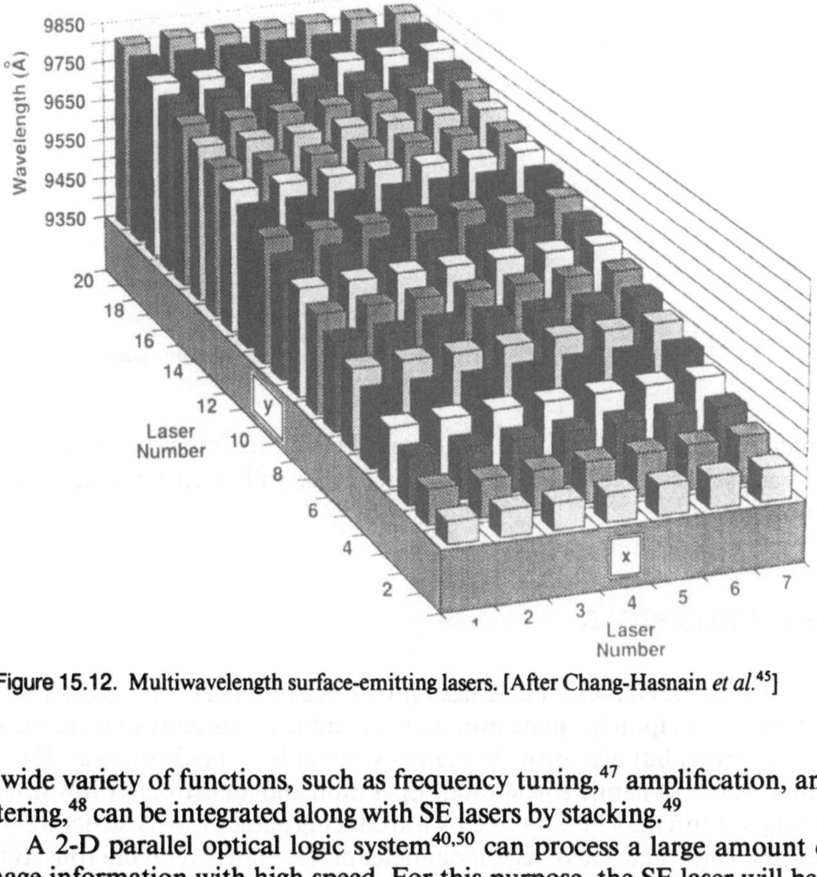

Figure 15.12. Multiwavelength surface-emitting lasers. [After Chang-Hasnain *et al.*[45]]

A wide variety of functions, such as frequency tuning,[47] amplification, and filtering,[48] can be integrated along with SE lasers by stacking.[49]

A 2-D parallel optical logic system[40,50] can process a large amount of image information with high speed. For this purpose, the SE laser will be a key device. Its technical aspects and future prospects are detailed in a textbook by Iga and Koyama.[51] Vertical optical interconnection of LSI chips and

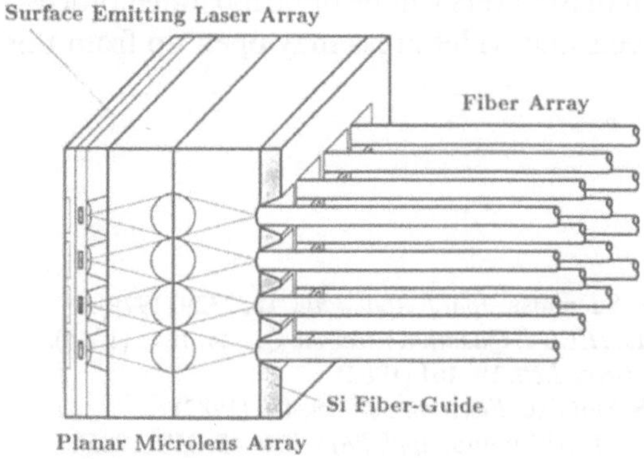

Figure 15.13. Idea for a parallel lightwave subsystem using a surface-emitting laser array and a planar microlens array.

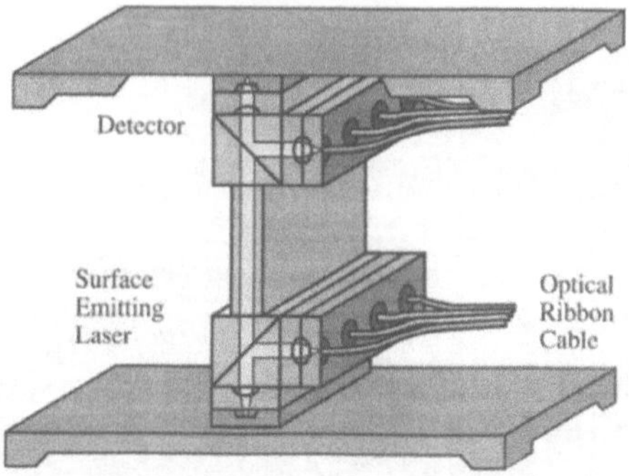

Figure 15.14. Optical interconnection using surface-emitting lasers.

circuit boards as shown in Fig. 15.14 may be another interesting issue.[52] The 2-D arrayed configuration of SE lasers, thus, will open up a new era in optoelectronics.

15.7. PROSPECTS

We have seen that a VCSE laser provides a number of advantages which not only are helpful for mass productivity and the possibility of forming a 2-D laser array, but also provide many excellent laser performances. For example, stable dynamic single-mode operation and an extremely low threshold ($I_{th} < 1$ mA) can be expected by introducing a microcavity structure with a cavity length and active region diameter of less than several microns. It has been recognized that the present performances of VCSE lasers are not limited by any essential problems, but only by technical ones. The development of basic semiconductor technologies, such as damage-free microfabrication process and sophisticated epitaxial growth, may be essential for the SE laser. The most interesting application of SE lasers will be optical parallel processing including optical interconnects and wider areas may open up from this kickoff.

REFERENCES

1. H. Soda, K. Iga, C. Kitahara, and Y. Suematsu, *Jpn. J. Appl. Phys.* **18,** 2329 (1979).
2. K. Iga, F. Koyama, and S. Kinoshita, *IEEE J. Quantum Electron.* **QE-24,** 1845 (1988).
3. Y. Motegi, H. Soda, and K. Iga, *Electron. Lett.* **18,** 461 (1982).
4. K. Iga, H. Soda, T. Terakado, and S. Shimizu, *Electron. Lett.* **18,** 457 (1982).
5. K. Iga, S. Ishikawa, S. Ohkouchi, and T. Nishimura, *Appl. Phys. Lett.* **45,** 348 (1985).

6. S. Uchiyama and K. Iga, *IEEE J. Quantum Electron.* **QE-20,** 1117 (1984); *Electron. Lett.* **21,** 162 (1985).
7. S. Uchiyama and K. Iga, *Trans. IEICE Jpn.* **E69,** 587 (1986).
8. S. Kinoshita, T. Sakaguchi, T. Odagawa, and K. Iga, *Jpn. J. Appl. Phys.* **26,** 410 (1987).
9. S. Uchiyama, Y. Ohmae, S. Shimizu, and K. Iga, *IEEE/OSA Lightwave Tech.* **LT-4,** 846 (July 1986).
10. K. Iga, S. Kinoshita, and F. Koyama, *Electron. Lett.* **23,** 134 (1987).
11. S. Kinoshita and K. Iga, *IEEE J. Quantum Electron.* **QE-23,** 882 (1987).
12. F. Koyama, S. Kinoshita, and K. Iga, *Trans. IEICE Jpn.* **E71,** 1089 (1988).
13. F. Koyama, S. Kinoshita, and K. Iga, *Appl. Phys. Lett.* **55,** 221 (July 1989).
14. A. Chailertvanitkul, K. Iga, and K. Moriki, *Electron. Lett.* **21,** 303 (1985).
15. T. Sakaguchi, F. Koyama, and K. Iga, *Electron. Lett.* **24,** 928 (1988).
16. M. Ogura, T. Hata, and T. Yao, *Jpn. J. Appl. Phys.* **23,** L512 (1984).
17. S. Uchiyama, K. Iga, and Y. Kokubun, *12th Eur. Conf. Opt. Commun.*, p. 37 (Sept. 1986).
18. H. Kawasaki, F. Koyama, and K. Iga, *Jpn. J. Appl. Phys.* **27,** 1548 (1988).
19. T. Baba, M. Matsuoka, F. Koyama, and K. Iga, *Optoelectronics Conference* (Chiba) **16C2-3,** 160 (1992).
20. M. Oshikiri, H. Kawasaki, F. Koyama, and K. Iga, *IEEE Photon. Tech. Lett.* **1,** 11 (Jan. 1989).
21. M. Oshikiri, F. Koyama, and K. Iga, *Electron. Lett.* **27,** 2038 (Oct. 1991).
22. Y. Imajo, A. Kasukawa, A. Kashiwa, and H. Okamoto, *Jpn. J. Appl. Phys.* **29,** L1130 (1990).
23. H. Wada, D. I. Babic, D. L. Crawford, J. J. Dudley, J. E. Bowers, E. Hu, and J. L. Merz, *Device Research Conference,* Post Deadline Paper, IIIA-8 (1991).
24. M. Tanobe, F. Koyama, and K. Iga, *Electron. Lett.* **25,** 1444 (Oct. 1989).
25. M. Shimizu, F. Koyama, and K. Iga, *Jpn. J. Appl. Phys.* **27,** 1774 (1988).
26. K. Tai, G. Hasnain, J. D. Wynn, R. J. Fischer, Y. H. Wang, B. Weir, J. Gamelin, and A. Y. Cho, *Electron. Lett.* **26,** 1628 (Sept. 1990).
27. A. Ibaraki, K. Kawashima, K. Furusawa, T. Ishikawa, T. Yamaguchi, and T. Niina, *Jpn. J. Appl. Phys.* **28,** L667 (1989).
28. D. Botez, L. M. Zinkiewicz, T. J. Roth, L. J. Mawst, and G. Peterson, *IEEE Photon. Tech. Lett.* **1,** 205 (1989).
29. M. Shimada, T. Asaka, Y. Yamasaki, H. Iwano, M. Ogura, and S. Mukai, *Appl. Phys. Lett.* **57,** 1289 (Sept. 1990).
30. Y. H. Lee, B. Tell, K. Brown-Goebeleer, and J. L. Jewell, *48th Device Research Conference,* Post Deadline Paper, VB-1 (June 1990).
31. K. Tai, G. Hasnain, J. D. Wynn, R. J. Fischer, Y. H. Wang, B. Weir, J. Gamelin, and A. Y. Cho, *Electron. Lett.* **26,** 1628 (1990).
32. M. C. Wu, M. Ogura, W. Hsin, J. R. Whinnery, and S. Wang, *Conference on Lasers and Electro-Optics,* WG4 (1987).
33. J. L. Jewell, A. Scherer, S. L. McCall, Y. H. Lee, S. J. Walker, J. Harbison, and L. T. Florez, *Electron. Lett.* **25,** 1123 (Aug. 1989).
34. R. S. Geels and L. A. Coldren, *48th Device Research Conference,* VIIIA-1 (June 1990).
35. T. Tamanuki, F. Koyama, and K. Iga, *Jpn. J. Appl. Phys.* **30,** L593 (1991).
36. T. Kobayashi, Y. Morimoto, and T. Sueta, *Natl. To Meet. P. Radiat. Sci.,* No. RS85-06 (1985).
37. Y. Yamamoto, S. Machida, K. Igeta, and Y. Horikoshi, *XVI Int. Conf. Quantum Electron.* WB-2, p. 27, 1991 (1988).
38. T. Baba, T. Hamano, F. Koyama, and K. Iga, *IEEE J. Quantum Electron.* **QE-27,** 1347 (1991).

39. T. Baba, T. Hamano, F. Koyama, and K. Iga, *IEEE J. Quantum Electron.* **QE-28,** 1310 (1992).
40. T. Numai, M. Sugimoto, I. Ogura, H. Kosaka, and K. Kasahara, *Jpn. J. Appl. Phys.* **30,** L602 (1991).
41. T. Tamanuki, F. Koyama, and K. Iga, *Jpn. J. Appl. Phys.* **31,** 1810 (1992).
42. S. Uchiyama and K. Iga, *Electron. Lett.* **21,** 162 (Feb. 1985).
43. E. Ho, F. Koyama, and K. Iga, *MOC/GRIN'89,* J2, 242 (Oct. 1989).
44. M. Orenstein, E. Kapon, N. G. Stoffel, J. Harbison, L. T. Florez, and J. Wullert, *Appl. Phys. Lett.* **58,** 804 (Feb. 1991).
45. C. J. Chang-Hasnain, J. Harbison, C. E. Zah, L. T. Florez, and N. C. Andreadakis, *Quant. Electron.* **27,** 6 (1991).
46. M. W. Maeda, C. J. Chang-Hasnain, C. Lin, J. S. Patel, H. A. Johnson, and J. A. Walker, *IEEE Photon. Tech. Lett.* **3,** 268 (March 1991).
47. N. Yokouchi, F. Koyama, and K. Iga, *Trans. IEICE Jpn.* **E73,** 1473–1475 (Sept. 1990).
48. S. Kubota, F. Koyama, and K. Iga, *Jpn. J. Appl. Phys.* **31,** L175 (1992).
49. K. Iga, M. Oikawa, S. Misawa, J. Banno, and Y. Kokubun, *Appl. Opt.* **21,** 3456 (1982).
50. R. A. Morgan, K. Kojima, T. Mullally, G. D. Guth, M. W. Focht, R. E. Leibenguth, and M. Asom, *CLEO'92* (Anaheim), CPD20 (May 1992).
51. K. Iga and F. Koyama, *Surface Emitting Lasers* (in Japanese), Ohmsha, Tokyo (1990).
52. K. Iga, *Trans. IEICE Jpn.* **E75-A,** 12 (1992).
53. T. Wipiejewski, K. Panzlaff, E. Zeeb, and K. J. Ebeling, *18th Eur. Conf. Opt. Commun.*, Post Deadline Paper, PD-8 (Sept. 1992)

INDEX